Totally Bananas

TOTALLY BANANAS

the funny fruit in American history and culture

DENNIS N. FOX

Library of Congress Number: 2002093929
ISBN: Hardcover 1-4010-7135-X
Softcover 1-4010-7134-1

This book was printed in the United States of America.

To order additional copies of this book, contact:
Xlibris Corporation
1-888-795-4274
www.Xlibris.com
Orders@Xlibris.com

16223-FOX

CONTENTS

For Lois, who inspired the idea, and
Marianne, who wholeheartedly embraced it.

CHAPTER 1

The Ubiquitous Banana

This handsome fellow learned to skate
By methods wild and queer;
He struck a fresh banana peel
And skated on his ear.
Anonymous

The banana is a funny fruit, the only fruit with a bright yellow "Have a Nice Day" permanent smile. Everyone it seems loves them, but we don't take them too seriously. Bananas are the world's most widely consumed fruit. A record 11 million tons of bananas were consumed in 1995. On average Americans by themselves were eating 28.6 pounds per person in 1998, an increase of thirty-eight percent since 1980. By comparison apple consumption remained stuck at 19.2 pounds per person during the same period.[1] Yet the humble elongated yellow fruit is the Rodney Dangerfield of fruits; it just "doesn't get no respect" compared to the American icon status accorded the polished spherical red neighbor in the produce department. No one ever said "hot dogs, banana cream pie, and Chevrolet," and no city has laid claim to the title, "The Big Banana." But it must have been the banana that was smiling when one writer observed that the Red Delicious apple was the "top banana" of the apple world.[2]

Perhaps we would all give the funny fruit more respect if we recognized that we are related to the banana. According to Robert May, an evolutionary biologist and president of the Royal Society of Great Britain, bananas share half their genes with humans. May noted, "This is a fact more evident in some of my acquaintances than others."[3]

Bananas are almost always funny. One can hardly get through the day without at least once reading or hearing that someone went "bananas," which is to say that the person was wildly enthusiastic if not mildly crazy. The world of business, politics and entertainment is filled with"bananas." Like early comedians of burlesque or vaudeville, Jay Leno and David Letterman vie to be the "top banana" of late night television. No one would deny that Bill Gates is the "top banana" of the cyberworld. Political analysts say that Al Gore was a great "second banana," although many would question his comedic skills. Perhaps the best known all-time second banana of entertainment is Ed McMahon who seemed perfectly suited to his role of foil to the *Tonight Show's* star, Johnny Carson. "When I hooked up with Carson . . . . that's what he was looking for," says McMahon. "A second banana."[4] McMahon played the role to perfection.

The language is replete with banana slang. If someone tells you a nonsensical tall-tale, it's "banana oil." If you're stupid enough to believe them, you are a "banana head." A "banana" in a hospital, is a person likely suffering from a case of jaundice. If someone speaks of your "banana" they could have one of two body appendages in mind. One is the facial characteristic that earned Jimmy Durante the nickname "Schnozzola;" the other relates to Freudian psychology and is the basis of the banana's reputation as a phallic symbol.

Thank goodness for banana humor. It can even add levity to a serious situation. Take the case of AIDS patient Jeff Getty, the 38-year-old San Francisco resident who in late 1995 received a risky infusion of baboon bone marrow that doctors hoped would save his life. Baboon cells are naturally resistant to the AIDS virus. Doctors theorized that a marrow transplant of baboon cells would improve Getty's immune system. With an uncertain prognosis, Getty held a news conference as he left the hospital three weeks after the transplant. He made special note that his new status as part baboon had led to "a lot of banana jokes."[5]

There are many uses for a banana peel although eating them is not a recommended use, and you had better not drop one on the sidewalk in Council Bluffs, Iowa where it is against the law. Banana peels got one Frank Smith of Philadelphia a six month jail sentence in 1921. Smith was arrested at the offices of the Public Service Railway Company in Camden, New Jersey when he presented a claim for injury suffered on a trolley car when he slipped on a banana peel. He insisted that his claim be settled out of court. An investigation disclosed that Mr. Smith had filed similar claims in other parts of the country. Smith confessed that he worked with accomplices who preceded him on the trolley and dropped the peels.[6]

When the seven justices of the Florida Supreme Court aren't deciding the outcome of U. S. Presidential elections, they sometimes wrestle with slippery food cases as they did in 2000 with two cases involving bananas. Florida seems to have a body of vegetable law arising from slip and fall cases. Previous slip-and-fall cases centered on collard green leaves, mashed potato, frozen peas, and sauerkraut. The Supreme Court cases in 2000 concerned a woman who slipped on "a brown, rotten banana peel" and a separate case of a woman who fell because of a "small piece of slightly discolored banana." The key issue in both cases was how long the banana was on the floor. If it was on the floor for only a few minutes, it would have been unreasonable to expect store employees to have cleaned it up. But if it was a longer period of time, the supermarket might be negligent. Lower court judges had thrown out the cases ruling that the color of the banana alone did not establish enough of a case against the store. The Supreme Court was considering whether the judges had made the right decision or if juries should be allowed to hear the cases.[7] Either way, someone would most likely appeal the case.

Patti Hagan, an urban gardener in Brooklyn, developed a more constructive use for banana skins—feeding her roses. She realized that banana skins rot quickly and provide considerable quantities of calcium, magnesium, sulfur, phosphates, sodium and silica. She collected "castoff banana epidermises" on Flatbush Avenue after teaching her retriever, Mojo, to point for them and arranged with the local Palestinian grocer to dispose of his blackened bananas.

She has a three-fold technique. Banana-skin fillets are laid flat at the base of each rose. Some blackened bananas are buried one banana per rose, and chopped banana skins are fermented for fourteen days in a sealed jar of water. Her roses are now "going bananas" in a paroxysm of blooms.[8]

More often the banana peel has been the cause of a comedian's pratfall in all sorts of entertainment. But did anyone in real life ever experience their undoing at the hands of the slippery peels? Allegedly, one Anna Hopwell of Enosberg, Vermont met her end in just this way. Her epitaph reads,

Here lies the body of our Anna
Done to death by a banana
It wasn't the fruit that laid her low
But the skin of the thing that made her go.[9]

Another person who had an unpleasant encounter with a banana peel didn't die from the experience, but his life was certainly changed. Trevor Hellems was an artist, teacher, philosopher and world traveler, perhaps best described as a combination of Plato and Benny Hill. He spent many years deep in the heart of banana country, India and southeast Asia, and his banana anecdotes were numerous. In India, the "banana leaf lunch" is quite common. Restaurants, or more appropriately proletarian lunchrooms, frequented by the locals use dried or sometimes fresh banana leaves rather than traditional plates. Young people rush among the packed-in diners dispensing copious quantities of rice and vegetable curries. One eats with only the right hand as in many eastern cultures the left hand is reserved for other "bodily functions." The locals regard the use of the hand and the disposable banana leaves as more sanitary than Western utensils. Hellems agreed. "You feel more secure than with a plate." With plates, "they just rinse them in a big pail of water, jiggle them off, shake them and put down the plate. Voilà, there you are. Eat!"

In most of the cities of South Asia there are dozens and dozens of street vendors selling crisp, tasty fried bananas. Hellems also noted that when traveling by train, bus or any public conveyance which has an attendant, they are quite happy if you leave them a bag of bananas as a tip. "They are terribly cheap; you get a bag of bananas for what is the equivalent of ten cents. We use them in bus travel because they are so available. It was there, your bag, and you knew where it was. You stop at lousy restaurants. They are so filthy, and you are a bit reluctant to go in and eat. But with bananas you could sit there eating a banana, and it does stave off hunger."

"Anyway," according to Hellems, "one of the things that is common in our culture is the flippant retort to the question of how did you hurt yourself? You usually flippantly reply, Oh, I slipped on a banana peeling. And I did slip on a banana peeling, and it just about changed the course of my life." It was during a vacation in Tangier, Morocco about thirty years ago. It was twilight, and Hellems was late for an appointment. "Twilight, and right in the traditional, classical position at the first step on a flight of stone steps going down to the lower city was a banana peeling which I didn't see until I was well up in the air twirling around coming down on my arm extended behind me. So it almost broke the arm right off the shoulder. There was pain which I will remember until the end of time." Passers-by were quick to help. "You know when anyone falls down, the first reaction is to get them on their feet. Everything is going to be all right if you get them up, get them on their feet." Hellems remembered saying, "No, no don't touch me, leave me alone." Finally, he was helped to his feet, placed in a cab and returned to his hotel.

Aided by his friends but in excruciating pain, Hellems set out in search of a hospital. They visited the French, the English, and the Spanish hospitals, a tripartite arrangement left over from colonial days. It was Friday night, and none of the hospitals was accepting patients. "Well we finally did get into the Moroccan hospital, and the young intern came. He turned out to be the intern, but I thought he was the teenage door attendant." He looked at the arm and announced, "Well this is an emergency, come back Monday and we will take x-rays." The young intern applied a plaster and a wrap almost "half the size of a bed sheet." Not satisfied with this solution, Hellems and his friends made inquiries and discovered that there was an old English physician who for many years had operated a charitable clinic in Tangier. The English doctor had an x-ray machine and agreed to open up as a courtesy to fellow Europeans. "He was able to take an x-ray, not too good an x-ray, but a good enough x-ray for him to say, 'get out of Morocco as fast as you can.'" The doctor gave Hellems some "terrific high potency stuff that knocked you out on your feet" and advised that the injured party take a plane the first thing in the morning to the American hospital in Madrid.

Trevor Hellems in Kathmandu, Nepal in 1988, far from his unfortunate encounter with the Moroccan banana peel. (Courtesy of J. and M.A. Farrell)

Hellems and his friends decided that since they were headquartered in Washington, D.C. they would take a plane there. Thus while heavily sedated and with an arm swollen to the "proportions of a watermelon," Hellems and friends arrived in Washington on the second morning after the accident. They went directly to a hospital where they encountered, as Hellems recalled, "two young Iranian interns who hated each other. They each had a theory about what should be done. One of them said I would be put in a sling-like yoke with the weight of the arm hung around my neck to keep a tension. The other one had a theory that it should be operated on and physically reattached. So then they had this big meeting with all the staff about what to do with this person and his arm." The intern that proposed to place the arm in a sling around the neck won out. "But it was just dreadful because of the anguish of having this awful yoke hung around your neck." Later while riding on the elevator Hellems again met the intern whose treatment theory had not prevailed. The intern said, "I'm sneaking you into the hospital tomorrow. You come back and go to the back entrance and we will be there to put you up in a room." Hellems agreed. The next morning he went back to the hospital at the back entrance, and "sure enough they had their little espionage set up. They whisked me into the hospital and up to my room, and nobody knew I was there." The intern arranged for surgery which was performed. With an obvious tone of glee Hellems recalled, "When the other Iranian intern found out I was there, oh he was mad. He had been tricked. So the other one that had finally achieved the coup was quite pleased with himself."

Shortly thereafter, Hellems was scheduled to report to work in a new teaching position in the Washington public schools. Afraid that if he did not report he would lose his position, "that morning I got up and I had this huge plaster of Paris cast all over the front of me. I couldn't wear a shirt or a tie or anything. It was just this huge plaster cast. I went to work to establish the fact that I was there. So after I got signed in, I went back home and was out for a couple of weeks." When he finally returned, "for most of the semester I was all casted up, and it was all for slipping on a banana peeling. So I don't even know today how I can tolerate being in the same room with a banana although I foolishly do rather like bananas." Hellems never regained full motion in his arm and for the remainder of his life carried the steel shaft inserted during surgery. Despite it all, Hellems said, "I've made my peace with the banana. Just one human being against billions of bananas."[10]

People are not alone in slipping on bananas. In February 1942 during the height of World War II a new freighter, the *Cape Romano*, was launched by the Pennsylvania Shipyards in Beaumont, Texas. Bananas were employed to ease the new ship's passage down the ways. The banana that launched a thousand ships? Not quite, but, shipyards had experimented with bananas for such a purpose for a year.[11] One can only speculate that bananas in short supply were more expendable than grease derived from precious petroleum.

Bananas are frequently the subject matter of proverbs and pithy sayings. Green banana references seem quite common in the sports world. When questioned about his plans after the 1995 British Open at St. Andrews, 65-year-old Arnold Palmer was non committal. "When you get to my age, you don't even buy green bananas, anymore." The concept isn't unique to mature golfers. A coach in the Canadian Football League was asked about job security for coaches in the league. The reply was, "Not many people in Canadian football buy green bananas."[12] While green bananas may not be very popular with the older set, ripe bananas appear to be popular with the youngsters. The young golf phenom, Tiger Woods, was seen strolling down the fairways at a 1996 PGA event with banana in hand thus setting off speculation that he may have been the first pro to do so. Nevertheless, athletes shouldn't worry, green bananas still ripen at a rate faster than athletes age.

As one might expect, the banana is involved with myth and superstition, particularly among fishermen. During World War II steamers carrying the fruit seemed to be popular targets for German U-boats. It was during that time that the banana's poor reputation among seafarers took root. Since then, the banana has become a harbinger of bad luck among yachtsmen. Fishermen are thus warned; if you expect the fish to bite, leave the bananas at home.[13] On the other hand, the banana can be a good luck charm. Some people wish upon the banana. Make your wish; then eat your fruit or, better yet, cut a slice from the stem end of the fruit while making a wish. Either way, if you find a Y-shaped mark at the end , your wish will come true.[14] Considering the banana's natural configuration, it's a "can't miss" proposition.

Hawaii, the only state in the United States to produce a significant quantity of bananas, has numerous banana-related myths. On the large island, Hawaii, one myth says if you travel to the Ka-u district and visit the valley above Hi'ilea, you will find a huge boulder of lava, the encrusted remains of Kumanna, a forest rain god who is said to have been a banana planter at a marshy spot in the valley. One day Pele, the fire goddess of the volcano, came to Kumanna in the shape of an old woman. Kumanna refused to share his bananas with her. An incensed Pele first sent cold against him; then, as he sat doubled up with this hands pressed against his face trying to keep warm, she overwhelmed him with molten lava—thus the boulder.[15] The morale of the story? When in Hawaii, share your bananas.

An authentic and reliable Hawaiian source reports that in traditional Hawaiian society before the arrival of Europeans females were forbidden to eat bananas. It seems that some foods were classified as "male" foods while others were female. The banana's basic anatomy clearly placed it in the "male" category and thus off-limits for female consumption.

Field workers in anthropology have recorded some bits of banana folk advice. In Ohio you might hear, "Man is like a banana; when he leaves the bunch, he gets skinned." And from New York comes, "If you want roasted bananas, you must burn your fingers first."[16] One gets the impression that they are not talking about food preparation.

Bananas are ubiquitous in American gastronomy. Perhaps the banana split is the most famous America food that is prepared from a banana. Ice cream in one form or another goes back at least as far as the Greeks and Romans, but the Americans made it an everyday staple. An ice cream machine for home use appeared around 1846. Commercial ice cream factories were operating by the 1850's, and ice cream parlors were soon found in American cities. No one knows with certainty when the ice cream sundae made its appearance, but Americans were enjoying the syrup and fruit-laden concoctions by the end of the nineteenth century. The introduction of the banana into the American diet near the end of the century opened new dessert opportunities. The fruit caught on fast as the foundation for a sundae. Elongated dishes were manufactured to accommodate the fruit and several scoops of ice cream which could each be adorned with luscious toppings.[17] While some might recognize Strickler's Drug Store in Latrobe, Pennsylvania as the originator of the banana split in 1904, the confection's birthplace is not certain. A Dispenser's Formulary in 1915 called the banana split the first "fancy fountain dessert" to win favor. Later, with fewer soda fountains and a greater health consciousness, it lost some of its appeal. Selinsgove, Pennsylvania may be an exception. On April 30, 1988 the citizens of that city created the longest banana split ever, 4.55 miles.[18]

Nation's Restaurant News reported in 1994 that bananas had replaced grapefruit as America's most popular breakfast fruit and had become one of the most popular fruits for restaurant dessert creations. As an easily prepared, relatively cheap and low-moisture fruit, chefs can use bananas in a great variety of ways. The banana flavor works well with chocolate, cream, caramel, and nuts as well as spices such as cinnamon and nutmeg. Greater interest in Caribbean and Pacific Rim cooking has also been a factor. Exotics such as coconut Bavarian with sugar-glazed banana and lime leaf sorbet or banana cream pie with banana crust, caramel sauce and chocolate shavings are appearing on American menus.[19]

Bananas are making more frequent appearances as flavorings or married to other ingredients in food and drink. One can "go bananas" in the food store. A supermarket tour reveals Orange-Pineapple-Banana juice, Strawberry-Banana chunky fruit nonfat yogurt and Banana Nut cream cheese. You can buy "Banana Vanilla Inclination" fruit drink, "Banana Raspberry" flavored water and "Banana Nut Cappuccino."

Preparing a banana split at a Southington, Connecticut soda fountain in 1942 (Courtesy of Library of Congress, LC-USW 3-042106-E)

Some foods named banana aren't even bananas. There is a banana pepper which is the shape and color of a banana. It is relatively mild and sweet, but it can easily be confused with the Hungarian wax pepper which is very hot. And if you tire of the typical supermarket potatoes, you might want to try colorful fingerlings like the Russian Banana potato. They are very tasty, but you will probably have to grow your own.

Of course there is the matter of how to properly eat a banana. Certainly the method employed by the winner of Estonia's first banana-eating contest in 1997 is not recommended. Mait Lepik downed ten bananas in three minutes which may not seem such a great feat except that he did not peel them. His prize was a free trip to the Canary Islands where presumably he could "pig out" on bananas.

How you eat bananas depends largely on where you eat them. Judith Martin, the well known Miss Manners, says in her *Guide to Excruciatingly Correct Behavior* that the banana as a "fruit occupies the place in the food world that the ingenue does in society. That is, it is usually fresh . . . . and although welcome anywhere for its charm and simplicity, it requires more complicated treatment when going about socially than it does when just hanging around the house." Miss Manners also notes that "For those who want to eat efficiently, God made the banana, complete with its own color-coordinated carrying case." Her instruction for eating the fruit informally is to peel it gradually using the bottom part as a holder. In more formal settings she recommends stripping the peel entirely away, cutting slices, and eating with a fork. However, she notes that "eating a banana with a knife and fork is almost as funny to spectators as slipping on a banana peel."[21]

Letitia Baldridge in the *Amy Vanderbilt Complete Book of Etiquette* notes the European custom of serving fresh fruit as a final course. In such company she advises eating a banana as the Europeans do, "their fingers never touching the fruit, their knives and forks doing the peeling, sectioning, and everything but the eating." At an America table Baldridge would use Miss Manners' method with the notable exception that a fruit knife and a fruit fork are required.[22] Both guardians of proper behavior would be appalled by a primitive tribe living on the island of Borneo whose table manners are worse than monkeys. When eating bananas, they don't peel them; they eat them whole.

Don't even think that you know the proper way to peel a banana because there may be more unorthodox habits than you would think. A prominent mathematical biologist—an unorthodox field to begin with—when in graduate school was known to always split his bananas down the middle and scoop out the seeds, which he discarded. He was once heard to remark, "I hate the banana pits!"[23]

Those who have considered the banana merely a food, haven't considered the magical possibilities of the fruit. A writer employed by the WPA South Carolina Writers' in 1939 recorded the story of Mrs. Josie Jones, a dressmaker, of Charleston. Mrs. Jones recalled her childhood in St. John's County, Florida. "Dad had about four hundred acres planted in sugar cane, potatoes, oranges and such. Then there was a big banana grove right back of the house. Playing lady was my favorite game. I'd be perfectly happy for hours at a time stringing the bright leaves from the sweet gum trees to make dresses for myself, or weaving the big palmetto leaves into hats for myself or little sisters. I loved to pretend that the banana trees were people, and I can hear myself right now saying, 'Howdy-doo, Mrs. Brown,' to the big trees, and 'Hello, children,' to the little ones." One can understand the flight of fantasy when we discover that Mrs. Jones, as one of the oldest children, had to constantly help her mother with the other children—all ten of them.[24]

Sometimes bananas can become entwined with religion. Take the case of "Crazy Denton," a name derived from the man's religious fervor, who in the 1890s peddled fruit and vegetables from a wagon in the Bushwick section of Brooklyn. The housewives who purchased from Mr. Denton received his personal assurance that the goods were the freshest and finest available. His lucky patrons also were instructed in the paths of rectitude which he believed would lead them to eternal joy in the hereafter. He questioned customers as to the state of their immortal souls in the same breath as he inquired about the condition of their larder. A Crazy Denton monologue might sound like this. "Good morning, Mrs. Jones! God bless you! I have the nicest spinach you ever saw. Yes, five cents a pound; all fresh. Thank you, ma'am! Are you saved? How about some nice wax beans, lady? Do you believe in Jesus? Those tomatoes are just off the vines. Remember, lady, Christ died to save sin—yes! sure I'll pick you out good ones. Come to Jesus, lady! I have some nice bananas. Are you washed in the blood of the Lamb? Where can you buy nicer green peas? Give your heart to the Lord! What's the matter with them apples? No! there ain't a rotten one among them! Those carrots are the best on the market . . . He died on the cross to save you and me . . . . four pounds for fifteen cents. The Lord be with you? Come to Christ!" Oh, the joys of banana salvation. Unfortunately, it was reported on more than one occasion that the self-appointed evangelist would abandon his fruits and vegetables for a a few days only to reappear with "an odor not unsuggestive of strong waters" on his breath.[25]

Strong waters were never associated with the thirtieth President of the United States, Calvin Coolidge, who aptly demonstrates the importance of the banana in presidential circles. During his presidency the taciturn Coolidge visited Longwood Gardens, the country estate of entrepreneur Pierre S. Du Pont, in Kennett Square, Pennsylvania. The President was taken by Mr. Du Pont himself on a tour of the 1,050 acres of outdoor gardens and 3.5 acres of greenhouses and conservatories. During the entire tour the President made no comment. The silence caused his host to wonder about the chief executive's health if not his mental state. As the tour was nearing its end in one of the conservatories, Silent Cal observed a large tropical plant with big leaves and familiar greenish-yellow fruit. The President paused, looked up, pondered a moment and spoke the only word he uttered through the entire tour. "BANANAS," he said.[26]

Bananas are important as well to more recent occupants of the White House. Appearing on the CNN show "Larry King Live" in April 1997, Hillary Rodham Clinton commented on how every word the first lady or President utters is noted. To illustrate she mentioned that on one late evening occasion the President decided he wanted a banana, but none could be found anywhere in the family quarters. The next day there were bananas everywhere in the White House. Presumably that also meant there were bananas in the President's small kitchen in the private quarters. At least that is what David Leopoulos, a childhood friend of Mr. Clinton's, thought when he spent a night in the Lincoln Bedroom. One of the perks of being the President's guest is that you get to rummage in the President's fridge for late night snacks. Dressed in ratty shorts, Leopoulos went in search of a banana late one night and stumbled upon the President deep in a wonkfest with the Vice President and a group of senators in the hallway. "I'm sure it was pretty important stuff," Mr. Leopoulos said, "I looked like this Greek refugee." The President waved him on, but the banana forgotten, Leopoulos headed back to bed.[27]

While the banana was not known to have ever gotten Mr. Clinton any votes, it can help immensely in Brazilian politics, especially if you are a monkey. In August 1997 Paulinho and Pipo, two chimpanzees, sought to replace Tiao the chimpanzee as the new symbol of Rio De Janeiro's zoo. Tiao was one of Rio's leading personalities, so popular in fact that in the 1988 mayoral elections he received thousands of write-in votes from an electorate that was not very impressed with the human candidates. When Tiao died in December 1996, the mayor ordered flags flown at half-staff. With anarchistic tendencies, Paulinho, age 12, made headlines after he pitched a banana at a top judicial official. His action won immediate favor with the public which was reminded of when temperamental Tiao threw water and sand at a former Rio mayor. The other contender, Pipo, age 24, was described as an ill-humored and conservative candidate. The campaign included marches, many pages of newspaper coverage and pollsters. Voting was at the zoo using ten electronic voting machines. Perhaps American politicians could use the tactics of primate politics to arouse public interest.[28] Let the bananas fly; they would be an improvement over mud.

Bananas may not be at the forefront of American politics, but they are in American business in a big way and not just with the companies that actually sell the edible banana. In some cases bananas can be used as currency if you are lucky. Tony Quirion of Bristol, Connecticut struck banana gold in May of 1997 when he read about a 1983 Cadillac in a flier from Chris' Auto Wholesalers that said the "First 10,000 Bananas Takes It." Taking the advertising at face value, Quirion contacted a Hartford fruit wholesaler and found that ten thousand bananas would cost about $1,100, which was less than half the cash price of the Cadillac. Chris Pio was surprised one Saturday morning when Quirion showed up with a mountain of bananas to claim the Caddy. "I opened my mouth, and someone brings in bananas. I had to hold myself to my word," said Pio.[29]

Using the word "banana" in a company name seemed all the rage in late twentieth century America. Sometimes the corporate evolution may make the name an anachronism. The national clothing retail chain, "Banana Republic" is such a case. The "Banana Republic Travel & Safari Clothing Company" was started in San Francisco in 1978 by Mel and Patricia Ziegler. They started with a $1,500 purchase of old Spanish paratrooper shirts which they quickly sold at a flea market.

Mel, a former reporter for the San Francisco Chronicle, hunted for old army surplus as he traveled the world. The Zieglers quickly discovered that there was a demand for the stuff and opened a store in Mill Valley. That was quickly followed by a catalog. By 1983 they were doing so well that they sought outside financing which came from the Gap, a successful California jeans company. The Zieglers retained creative control of the company. By 1985 there were two dozen Banana Republic stores in the United States, and over 12 million catalogs were mailed out. The stores resembled everything from a British officer's club in India to a jungle hunting lodge. Some of the mall store entrances looked like the misplaced film set for a Tarzan film. If you called the 800 number to order a $24 pith helmet—a genuine Bombay Bowler used by Her Majesty's forces in India—or Israeli fatigue shirts at $28, you would have been treated to a variety of rain-forest sounds while on hold.[30] By 1988 the "Out of Africa" look began to change both in the Banana Republic stores and in its advertising. The stores are no longer khaki kingdoms; today's four hundred stores are meticulously and tastefully designed and decorated— almost clothing museums with "solstice lighting." The Howler monkey background music is now jazzy—Judy Garland, "Putting on the Ritz." The only remnant of the safari experience is the Banana Republic label sewn in the clothes. The advertising is now more about attitude than fashion, and "free soul" couples are seen walking on the edge of the urban abyss.[31] Some might say it's the urban jungle now.

The Banana Republic catalog, discontinued in 1989, reappeared in 1998 and reflected the new image of casually luxurious fabrics in modern silhouettes. Now calling the 800 number connects you to a "style consultant" who can offer fashion advice or take your order for a $20 short-sleeved tee or a $398 leather peacoat—just the thing for the tropics.[32] But if you still feel the need for a World War II machete or pith helmet, you might try the Banana Bay Trading Company in Austin, Texas.

San Francisco seems to be a place where banana businesses begin. "Bananas at Large," a firm which in 1997 was selling several million dollars worth of guitars and equipment each year, was started there in 1972 by three guitar players. The "banana" part of the company name derived from the fact that all three founders were a bit wacky, again a nod to the banana's inherent humor. The "at large" came from the fact that the founders combed the countryside and pawn shops for guitars to sell in their story. There is something in a name after all.

Bananas are ubiquitous in business. If you thought you knew what a banana boat was, guess again. In the business world "Banana Boat" isn't a vessel on which an immigrant recently arrived; its a brand of sun care product. You might want to use "Banana Boat" if you visit a "Banana Bungalow," a chain of youth hostels in sunny climes such as San Diego—a bright yellow building only inches from the beach— and Miami. Even the Banana Bungalow in New York City has a sun deck on the top floor and a tropical theme lounge/kitchen. If you want to make a reservation at one of these fun-loving establishments, you might try the "Banana Pages" for the phone number. "Banana Pages" is a marketing name for neighborhood telephone directories published by NTD Publishing. The name is to differentiate from utility phone company directories. Guess the color of paper used.

Bananas also make cameo appearances in ads for other products. Jackie Chan, the acrobatic, martial-arts super-stunt man appeared in a 1997 television commercial set in Hong Kong. Chased by a bevy of Asian thugs through streets and markets, Jackie managed to find time to peel and eat a banana before escaping the clutches of the bad guys. What was he selling? Mountain Dew!

Charles Nelson Riley in the role of a teacher in an elementary classroom sang his way through a 1997 TV commercial for BIC Banana Ink Crayons. His banana costume was quite becoming, and his class of banana clad munchkins was adorable.

Two other television commercials of 1997 vintage either mocked bananas or elevated them to a pedestal depending on your interpretation. One ad for a termite exterminating company showed a perplexed puppy watching as a procession of the pesky little critters carry off family possessions including the dog food dish and a banana. Do termites eat bananas? The other ad was an ABC Television fall promotional piece which opened with the scene of a quivering molded red jello and a bunch of bananas. The only sound was a high pitched tone, presumably emanating from the jello. Bananas don't talk that way. After ten seconds of quivering, the tag line, set against a banana-yellow background, poses the question, "TV, what would you watch without it?" Advertising gurus were not quite sure what to make of it. Was it cynical self-parody? Was it creative exhaustion on the part of the Madison Avenue crowd?[33] The answer is obvious: back to basics—jello and bananas!

Robert Nelson, the author of *1,001 Ways to Reward Employees*, tells the story of where a banana is not just a banana but a corporate symbol. According to Nelson the story of the"Golden Banana Award" is part of the oral tradition at the Hewlett-Packard Company. The story is that one day a company engineer entered his manager's office with the news that he had solved a longstanding problem. The manager, appropriately impressed, looked around his desk for a fitting acknowledgement of the feat. He came up with the banana from his lunch which he offered as a congratulatory token to the engineer. Thus began the "Golden Banana Award;" or did it? Hewlett-Packard denies that any such award ever existed. Well, it's a good story nonetheless.[34]

In Snowmass Village, Colorado the Resort Association has discovered that bananas are good for business on the ski slopes. The annual "Banana Season" is a zany festival to welcome the arrival of spring. Among the events is a Banana Bonanza Hunt, a kind of Easter egg hunt for adults. Plastic bananas containing vouchers for prizes are hidden all over the mountain. Clues are given on the radio and on the boards at the bottom of ski lifts. During the Snowmass Ski Splash Banana Season Barbecue Bash, the "splash" is what happens as wackos ski off a ramp into a little pool while dressed in costumes—or not dressed at all. The contestants who do best have probably had too many "Funky Monkeys" or "Banana Smashes" in the Bartender's Banana Brawl Drink Contest.

What's in a name? It may not be a banana. B.A.N.A.N.A. is not an acronym for an organic vegetarian terrorist organization nor does it represent a Central American intelligence agency. It means Build Absolutely Nothing Anywhere Near Anybody, an organization that caused the city of Boulder, Colorado in 1993 to enact a temporary moratorium on any residential or commercial development. The idea was to prevent the eastern frontier of the Rocky Mountains from becoming one long polluted strip city.[35] Just the kind of environment to raise kids, but not bananas. And what possibly is a Chrome Banana? Well it's in the same category as Electric Prunes, Moby Grape or Iron Butterfly—a rock-and-roll band of the mid-60's. If you never heard of them, it is because this group of three seventh-graders never quite got past the elementary school parties, a few high school dances, the Lion's Club Minstrel show and a few talent shows in the vicinity of Lewistown, Pennsylvania. Still the Chrome Banana, they now perform only for class reunions.[36]

Another thing which masquerades as a banana is ariolomax dolichophalus, better known as the banana slug. This bright yellow, slimy, shell-less creature is a tropod mollusk and is native to the American Northwest where it can feed on fungus and debris in the damp surroundings of the California Redwood. It can grow to a length of eight inches— ten inches if it's lucky. Its color can range anywhere from black to white but is very often a bright yellow with dark spots—not unlike a very ripe banana. In 1986 this humble creature became part of a slugfest on the campus of the University of California at Santa Cruz when the students demanded that this unofficial mascot be made the official symbol of the school's athletic teams. Chancellor Robert L. Sinsheimer selected the sea lion as "a mascot with spirit and vigor." When he blocked a student referendum on the issue, the students started a massive grass-roots campaign for the lowly creature. Slug partisans believed that the slug captured the spirit of their unorthodox school which has no athletic scholarships, fraternities, sororities and gives grades by request. Said Eric Satzman, the president of the Student Assembly, "the slug represents our uniqueness and our resolve to stay that way." The sea lion mustered little support; one incensed alumnus complained that "all sea lions do is fornicate on the rocks while making grotesque sounds." In the face of overwhelming pro-slug support, President Sinsheimer named the Banana Slug the official mascot and noted in an official press release,"I would also suggest that it would be most desirable for our biological scientists to begin a program of genetic engineering research upon the slug, to improve the breed. The potential seems endless."[37]

While the bananas at UC-Santa Cruz may be slimy, "Bananas" at the University of Maine at Orono is furry. That's because "Bananas" is the name of the Black Bears' mascot. Legend has it that the first school mascot was an elephant which was "borrowed" by a student named Seldon and two accomplices from a Bangor clothing company. The pachyderm was ingeniously hidden under the stadium until the next football game. But the care and feeding of elephants being what it is, there must have been considerable relief when a tiny bear cub named "Jeff" was loaned to the University in 1914. Jeff made his first appearance at a football rally prior to a game with Colby. Jeff's antics caused the enthusiastic crowd to "go bananas." The football team overwhelmed Colby and Jeff's mascot performance was so inspired that the University's athletic teams have been known as the Maine Black Bears ever since.

No one knows what happened to Jeff after that first year, but a new bear was presented to the University of Maine in 1915. Art Smith, the football coach remembered the crowd's reaction to Jeff's first appearance and named the new bear "Bananas," a name that has stuck ever since. The athletic teams enjoyed a stellar year, and Bananas' fame spread. The U. S. Naval Academy even requested that the Maine Black Bear appear on its side of the field for the annual Army-Navy football game. The only problem with a live bear as a mascot is that bears hibernate in winter. A makeshift den was created in an old abandoned pump house. Bananas bedded down after the football season but was back in business for baseball.

The live bear tradition continued with a few interruptions until the death of "Cindy Bananas" in 1966. After that a Maine court decision outlawed the tradition. In 1969 a tradition of a human mascot began when Robert Smullen of Alpha Phi Omega fraternity proposed the idea and donned the first made-from-scratch bear suit. Today "Bananas" is a furry blue "Yogi-Bear" type character hugged by kids and cheered by Alumni.[38]

Bananas also seem to turn up in the legal system with some regularity. The nation's premier law enforcement agency, the FBI, has been known to use a little banana trickery to extract information from suspects. In 1996 the Bureau was investigating allegations that one Robert Kim had provided secret government documents to his native South Korea. Agents watching hours of surveillance video noticed that Kim had a special taste for bananas. When Kim was brought in for questioning, agents made sure that a basket of bananas was waiting for him. The purpose was to send Kim the message that they knew everything about him. Kim was later indicted. A little banana deception helped.

Some bananas and cocaine originate in the same part of the world, Columbia. Therefore, it should not be surprising that both items might be found in close proximity to one another. Such was the case in September 1990 when federal agents found 145 pounds of cocaine in a six foot length of eight inch plastic pipe bolted to the bottom of a banana boat in Bridgeport, Connecticut. Worth about $2 million it was the largest single seizure in Connecticut history. The ship had started its trip in Ecuador with bananas and plantains, but it had also stopped in Turbo, Colombia.[39] Less than a year later about twenty-five pounds of ninety-nine percent pure high grade cocaine was found packed in ten individual pouches in banana boxes in a Florida supermarket. Similar amounts were shortly found in another Florida store and one in Indiana. The cocaine was traced back to ships which had unloaded in Bridgeport and in Tampa, Florida. Four stowaways had been found on the Tampa ship and sent back to Colombia before the cocaine turned up in the supermarkets. Authorities believed that the cocaine smugglers were often stowaways.[40]

The banana has long suffered from its libido arousing shape which Viennese psychologists would describe as a "phallic symbol." Banana defenders work hard to keep the fruit virginal. When a PBS program in 1987 used a banana to demonstrate the use of a condom in a program on AIDS prevention, Robert M. Moore, then President of the International Banana Association, launched a barrage of protest. In a letter which appeared in *Harper's Magazine* he took exception to the "unsavory association" and pointed out that the "banana is an important product and deserves to be treated with respect and consideration." He went on to defend the banana's image as a healthful and nutritious product. It is the most extensively consumed fruit in the United States and is purchased by ninety-eight percent of American households.[41] In a sense he was calling it an American icon, which it is. It's just its anatomy which keeps causing problems. In the 1970s a publicist for a major banana company investigated rumors that he had heard that there was a women's prison in the Southwest that did not permit bananas within its walls. His investigation proved the rumor true. Prison authorities stated that they had no axe to grind with bananas per se—inmates were allowed banana puree—it's "just not, well, you know, just not whole bananas."[42]

When PBS used the banana in its demonstration in 1997, it might have been hauled into court under one of the food defamation laws passed by dozens of state legislatures in the nineties. Opponents refer to these laws as "veggie libel" or "banana bills." The laws were designed to make it easier for growers and ranchers to recover damages from those who allege health risks in their products. The problem arose in 1989 when the CBS "60 Minutes" program ran a piece that said that Alar, a chemical used in apple ripening, increased the risk of cancer. The Washington apple growers sued claiming that the program had defamed their crop. The suit was dismissed because the statements were not proven false and a food could not be defamed. Growers then pushed for passage of the "banana bills."

The movement seemed to culminate in 1997 when Oprah Winfrey was sued for knocking hamburgers. After a guest on her program claimed that "Mad Cow Disease" could plague the U. S. beef industry, Winfrey said, "It has just stopped me cold from eating another burger!" Texas cattle ranchers claimed the remark caused millions of dollars in damages to their industry. The ranchers eventually lost their case but not before a group of incensed emu ranchers filed suit against Honda for a funny commercial that mocked a young rancher trying to raise the Australian flightless birds.[43] While the civil libertarians battle the food disparagement bills on First Amendment grounds, it would be better not to castigate cauliflower or bad-mouth broccoli. And please, show the banana a little respect.

Of course no discussion of the legal system and the banana would be complete without mention of O.J. Simpson. In April 1998 after a day long interview in Los Angeles with British talk show host Ruby Wax, Simpson lunged at the interviewer and made repeated stabbing motions. The weapon of choice was a banana. The bizarre moment ended with Simpson leering in extreme closeup into the camera.[44] You can't make this up.

There is positive news about the banana. It can be used in family planning—at least for elephants. In 1992 the Kenya Wildlife Service was discovering that elephants as a protected species were prospering, so much so that they were doing a lot of damage. The Wildlife service was looking for a way to reduce the herd. Controlled shooting was not a solution that tourists would find attractive. Family planning seemed like the solution; keep the cows from giving birth when young and get them to calve less frequently. Hormonal contraception—an elephant version of the Norplant system—was considered. The problem was that the implant would have to be as big as a basketball. Unsightly to say the least. Abortion looked like a better solution. Enter the banana. It seemed that bananas spiked with RU-486 might do the job. Timing was essential. Given the doctored bananas too soon in the pregnancy, the cow would promptly conceive again. If given too late, the elephant would stand guard over the dead fetus or carry it around on her tusks for days. Such a sight would be unsettling to the other females, and the tourists would be horrified. Half way through the twenty-two month gestation period was the ideal time, and a number of methods were available to determine this without an ultrasound examination for elephants.[45] Yes, the banana is a versatile fruit.

There is just no escaping America's most popular fruit; bananas pop up everywhere. Recently, a woman and her elderly mother were observed in a supermarket checkout line. The woman had just about finished unloading both of their baskets when she announced, "Oh, I forgot to get one banana." "One? One banana! What for?" inquired the mother. "It's for my rabbit," came the reply. The observer could not let the moment pass and proceeded to inquire. "Yes, my rabbit just loves bananas," explained the younger woman, and she went on to point out that if she returned from grocery shopping without a banana, the little dear would be terribly upset. The observer wondered how an animal not noted for any great powers of perception would even know that its mistress had been grocery shopping, but with several *National Enquirer*, or *Globe*, or *Star*-crazed homemakers maneuvering their wheeled battering rams hard on his tail, the issue was not pursued. Later as the two women rolled their baskets off into the parking lot, the observer considered if he should have at least asked the banana bunny's name. Would you be willing to lay odds that if it wasn't "Fluffy," it had to be "Chiquita?"

CHAPTER 2

The Banana Family Tree

"It is evident that the banana, in every step of its history, has proven to be, in the phraseology of Cotton Mather, 'a special dispensation of Providence,'"
Dr. Herbert J. Spinden, 1926[46]

They were not supposed to be there; at least that's what many folks thought. But there they were. Bananas in all their glory. Bananas, twelve acres of them, in southern California. From the mid-1980s until 1998 motorists headed north from Ventura to Santa Barbara on Highway 101 could not but notice the thousands of banana plants nestled on a narrow shelf of land below a chaparral-covered three hundred foot high bluff on one side and the Pacific Ocean on the other. The fields of bananas, complete with a thatched-roof fruit stand, looked like a postcard from Hawaii. The setting was the Seaside Banana Gardens in La Conchita, not much more than a wide place in the Pacific Coast Highway.

Banana plants are not unknown in southern California. They are often seen in ornamental plantings, but the climate is not consistently warm enough for the plants to bear fruit. Almost all bananas are grown in tropical climates between 30 degrees North and South of the equator. All large commercial banana varieties are grown within the even narrower confines of 20 degrees on either side of the equator. The plant loves temperatures around 90°F with high humidity. At temperatures below about 55°F the plant stops growing. A frost is usually deadly, and southern California does get frosts. An average of four inches of rainfall per month and moist, well drained soil is required. Given these requirements you will find banana cultivation in Panama, Uganda and Indonesia but not in Canada, Sweden or Mongolia.

Bananas can be grown in colder climates but only with extraordinary artificial growing environments which are not economically efficient. If the United States was suitable for banana production, someone like Thomas Jefferson would probably have tried. Monticello's six acre fruit garden had eighteen varieties of apples, eleven varieties of pears, currants, figs, cherries, grapes, apricots, plums, almonds and thirty-eight varieties of peaches, but no bananas.[47]

Doug Richardson, the founder of the Seaside Banana Gardens, can't claim the fame of a Jefferson. But he is equally a horticulturist, and he had a dream. From the time that he was a geography major at the University of California-Santa Barbara Richardson knew he was interested in farming. He researched the possibilities of dozens of crops. "I really had a dream about getting a piece of land, planting it and growing something," he recalled. "All the crops I looked at were fruiting plants and typically fruit trees. I did a lot of research on each one, and I just couldn't afford to buy the land, make all the investments in the trees and the irrigation system and then wait what was typically five to eight years until you even reached a point where you were covering your costs."[48]

Richardson started out as a landscape contractor planting things for other people. He designed and installed plantings using mostly California native plant material including edible plants. "I was really fascinated with unusual edible plants," says Richardson. "We were setting our customers up with these fabulous edible gardens, and it really intrigued me to be just constantly researching new kinds of things that we could try." His search for new edible plants took him to nurseries in San Diego where he found a few varieties of banana plants. "I brought some of them back and planted some in my own yard, planted some in several different clients' jobs, and they were doing really well at our house in La Conchita."[49]

On a family outing to Santa Barbara Richardson's four young

daughters got ice cream cones, and their dad visited a used book store with piles of old *National Geographic* magazines. "I was very much interested in Mediterranean sub-tropical agriculture. So that anything I could find that treated with that, I got it. It just so happened that I laid ahold of this great article on the Canary Islands, and they spent several pages talking about the banana industry there. Here I had just put these bananas in at our place, and the similarity between La Conchita and the Canary Islands was just striking. I was a geography major, and I was looking at the terrain. It was really fascinating." Fascinating enough that Richardson launched some climate research.[50]

For four or five years Richardson kept high and low temperature records for La Conchita. What he discovered was a unique microclimate that did not experience a frost. The eastern bluffs isolated the coastal strip from cool night air from in-land mountains. The relatively balmy ocean breezes kept the air circulating and kept frost from settling in.

While he was doing his meteorological work, Richardson was also adding to his collection of banana plants in his front yard. Said Richardson, "We literally stripped all other vegetation off this little lot that we had at this house we were renting and stuffed it with bananas. I would come home from a day of landscaping, and I would kind of walk out into the yard and see how all of the banana plants were doing." It was a kind of hobby that went on for four or five years.

One day a friend gave him a banana variety brought from Hawaii. Richardson vividly remembers what happened next. "It grew so well and so quickly that within twelve months from the time that I planted it it was twenty-five feet tall. It had flowered and set this big, gorgeous bunch of bananas Three to three and a half months later the bananas were ripening and turning yellow, and they were absolutely the best bananas I ever tasted."[51] Richardson's gorgeous banana plant was the "Brazilian" or "pome," a variety grown principally in Hawaii for local consumption. It is said to have come to Hawaii in 1855 from the island of Java in Indonesia by way of Tahiti.[52] Says Richardson, "It is just this incredibly vigorous hardy banana. The bunches are not particularly large, twenty to fifty pounds. Depending on how severely you're pruning your clumps, how well you're feeding and fertilizing, and how good the climate is, the bananas are anywhere from three to four inches long up to six or seven inches long. That's the banana that actually propelled me into this business."[53] Richardson's experience in California, inspired by a *National Geographic* story on the Canary Islands and successfully growing a banana that crossed the Pacific from Indonesia, represents a symbolic joining of two globe-girdling migratory paths for the fruit we call the banana.

The first bananas in the New World arrived after 1492 as part of what is often called the "Columbian Exchange." When Europeans arrived in the Western hemisphere, they brought the hog, wheat and rice, and took back the potato, corn and the tomato.[54] Father Tomás de Berlanga brought the first banana plants to Hispaniola (now Haiti and the Dominican Republic) from the Canary Islands in 1516. They spread rapidly. Francisco Oviedo, inspector general in the West Indies for King Ferdinand, reported in 1525 in his *Natural History of the New World* that there were four thousand banana plants under cultivation on his own plantation and other larger plantations.[55] Other Spaniards during the sixteenth century observed that the banana and its relative the plantain were growing in the hot parts of New Spain. They commented that the fruit was not native and had been brought from Africa or the East Indies. By the early seventeenth century European travelers repeatedly mentioned that the plantain was cultivated by the Indians. Because the crop suited the New World environment and Indian needs, it diffused quickly leading some people to assume that the plant was indigenous to the New World. There is little credible evidence that the banana and plantain were anything other than immigrants to the New World.[56] Just as the horse and disease—also part of the "Columbian Exchange"—transformed the culture of the New World in the sixteenth century, the banana so transformed some of the Central American republics in the twentieth century that their economies are now heavily dependent on the fruit for their survival.[57]

Biologically, the banana is in the genus *Musa* of the family

Musaceae which also includes lilies, orchids and palms. The wild ancestors of most of today's cultivated bananas are found in Asia ranging from eastern India through Southeast Asia into the Malaysian peninsula and archipelago. The migration of the banana began there at some time in the prehistoric era. Two of the main characters in the story are wild banana species, *Musa acuminata* and *Musa bulbisiana*. Both are members of the section Emusa of the genus *Musa*. Most edible bananas today have evolved from these two wild bananas as hybrids known as triploids because they have three chromosome chains. The uneven number of chromosomes means that the plants are seedless and sterile. The two wild species developed seedless hybrids either by cross pollination or by a form of mutation known as sporting. When and how man became involved in this process is not precisely known. It is probable that human selection aided the development of varieties with tiny or nonexistent seeds.[58] Wild bananas with small hard seeds in the fruit would not have been an attractive food although other parts of the plant such as the corm (the root) and shoots may have been eaten.[59] The typical seedless banana that evolved through parthenocarpy (the development of fruit without pollination), would have been selected for replanting by our ancestors who knew a good thing when they saw it.

Archaeobotanists have searched for the origins of banana cultivation, but plant remains can disintegrate quickly in the tropics. Thus there is little primary evidence of Southeast Asian banana domestication in the form of plant remains in archaeological excavations.[60] The first recorded occurrence of specimens of wild banana in an archaeological context were recovered from Sri Lanka in 1983. The Bali-Lena cave site produced seeds tentatively identified as *Musa acuminata* that came from excavation levels that radiocarbon dating placed at 10,500–8,000 B.C. Today both human and animal inhabitants of jungle habitats use wild bananas. The seeds are inedible, but the tissue surrounding the seeds is sweet and edible. It is assumed that the earlier people of Bali-Lena used bananas in the same way, but they did not cultivate them.[61]

Until the 1980s archaeobotanists could determine the age of

recovered seeds only by association with larger items found together such as in the Bali-Lena site. Today both low and high-tech methods are used. Beginning in the early 1950s a low-tech system of seed recovery was developed. It was a flotation system. In simple terms excavated soil was dumped into water. Seeds, charcoal and small plant parts would float to the surface, but estimating the age of such small bits was a problem. To be reliable conventional radiocarbon dating usually required one to five grams of carbon. Small seeds would be completely destroyed in the process. In the late 1970s the development of radiocarbon dating by accelerator mass spectrometry made the dating of small seeds possible. Since the 1980s, the use of the scanning electron microscope has permitted the precise measurement of seed coats. Because domesticated species have thinner seed coats than their wild ancestors, it is now possible to more closely estimate when and where bananas were domesticated.[62]

The commercial bananas grown today have sterile seedless fruits, and such plants would not survive unless continuously planted by man. The man and banana relationship has gone through several phases to arrive at today's situation. Ten thousand years ago late Pleistocene man on the island of New Guinea probably manipulated the rain forest by opening the forest canopy to sunlight to an extent sufficient to exploit the nature stands of wild bananas found there.[63] It was a first step toward banana cultivation in which the plants might continue to exist as a breeding population without the interference of humans. But we can only say that the banana was domesticated when it became so genetically different from its wild ancestors that its very existence depended on human intervention in its life cycle. Once that happened, the domesticated banana became a cultural artifact.[64]

To better understand the process of domestication and migration of the banana a closer look at the plant itself is in order. The banana plant is often called a "tree," but it is actually a perennial herb—the Godzilla of the herb world. The plant is vegetatively propagated from a below ground bulb-like corm—known in the field as a rhizome—that sends out roots from its underside and stems from its top. This happens much in the same way that potatoes are propagated from pieces of potato tuber. The "trunk" or pseudostem of the banana plant is the overlapping bases of leaves wrapped tightly and, depending on the variety, may range from three to thirty feet tall. These leaves unfurl at the end to create the apparent leaves which may number as many as forty-four although some die off as the plant grows. A typical leaf may be nine feet long and two feet wide or even larger.

The number of leaves produced is an indication of how happy the plant is, and unhappy plants don't produce as many bananas. But banana leaves are a tough lot. Since the pseudostem is non-woody and ninety-three percent water, it is subject to easy blowdown by high winds. But nature has engineered a built-in defense. The veins in the leaves run straight from the mid-rib to the outer edge of the leaf. Thus the leaves can shred in high winds reducing the possibility of blowdown, and the veins can still function.

The true stem of the plant arises from the corm. As the inflorescence or flowering part of the plant ascends through the middle of the pseudostem, it is said to "bull." When it is first visible at the top, it "peeps." It has "shot" when it is fully emerged, an event that takes nine to fifteen months from the time the plant started growing. After the inflorescence has emerged, its weight causes it to droop, and it becomes pendent. At the end of this stem are the sterile male flowers protected by large red bracts. They are relatively useless except that they can make an interesting salad ingredient. Higher up on the stem are the groups of female flowers from which the seedless fruit (technically a berry) develop without fertilization. In a wild state the female flowers are pollinated in a normal fashion. The stem will usually have from seven to twelve double-rowed half spirals of fruits known as "hands." Each hand may have from twelve to twenty individual fruits known as "fingers." An average stem may have 150 individual fruits. These numbers vary depending on the variety of banana. A full stem of developed fruit can weigh as much as 100 pounds.

The flowering part of the banana plant ten days after it has "shot." (Courtesy of Library of Congress, LC-USZ62-86131)

A banana plant gets considerable attention while it is growing and especially when the fruit is developing. Leaves that are dead or touching the fruit are cut away. A perforated polyethylene bag is placed over the entire stem of fruit usually after all the hands have emerged. In some areas the interior of this plastic bag may be treated with an insecticide or a fungicide. This bagging is done to protect the fruit from insects, bats or scarring caused by leaf contact. A colored ribbon is placed around the end of the bag as a guide to the age at which the fruit will be harvested. The bag creates a kind of greenhouse effect that speeds fruit maturation by several days and can increase the weight slightly. When the stem is bagged, the male flower bud and a few small or incomplete hands are removed from the stem.[65]

As the stem of fruit develops, the plant can fall as a result of the increasing weight of fruit. To avoid this possibility plastic twine guys are tied to adjacent plants or to poles, or the plant is propped up by a wood or bamboo pole.

Fruit maturity is determined by an age and grade system. "Grade" is simply the diameter of the fruit as measured by calipers calibrated in thirty-seconds of an inch. Harvesting orders are given by grade and ribbon colors. Under normal circumstances workers will go through a field at least once a week to harvest the mature fruit.[66]

Harvesting is accomplished by two-man teams, a cutter, or "cortador," and a backer, or "cablero." The cutter uses calipers to measure fruit of the proper grade—typically one and a half inch diameter—on the middle finger of the outer row of the second hand. Using a long pole with a chisel-like knife, the cutter makes a shallow cut in the pseudostem so that the stem of fruit descends slowly to the padded shoulder of the backer. Rough handling of even green fruit will cause bruising. The cutter then severs the stem from the plant and cuts off the top of the plant while the backer carries the stem to a cable system or cart to remove the fruit from the field.[67] We will save the rest of the story of the banana's journey from field to market for later.

After the plant has borne its fruit it is usually cut down or part of the pseudostem is left to decay. Meanwhile, the rhizome has produced suckers which start the whole process over again. The entire group of stems that develop from one rhizome are known as a "mat." A typical "mat" may produce many suckers, but commercial producers seldom allow more that three suckers at various stages of development to grow at one time. Experienced banana men know which suckers to prune and which to leave. If not destroyed by disease or other natural disaster, a commercial field of banana plants may be kept in cultivation for up to twenty years before the field is replanted. Small patches of bananas in the tropics may be cultivated by farmers for fifty or sixty years.[68]

It is clear that bananas and plantains have been "bulling," "peeping," and "shooting" for thousands of years. Early historic accounts frequently do not clearly distinguish between the two edible members of the genus *Musa*. The common banana, sweet and usually eaten raw, and the plantain, more starchy and usually eaten cooked, are interchanged in early accounts. The earliest literary references are in Buddhist sacred books dating back to about 600 B.C. The banana plant is also depicted in sculpture in the ruins (175–100 B.C.) of ancient Buddhist temples in India.[69] Carolus Linnaeus, the Swedish father of botanical classification, assigned the non-Latin word *Musa* for the banana's genus. *Musa* comes from the Arabic word *Mouz* which was derived from the Sanskrit word *Moka* or *Moca*.[70] All of this argues that Europeans were aware of bananas growing in India.

When Alexander the Great led his army across the Indus River in his conquest of India in 327 B.C., he was impressed by the banana. Shortly thereafter in his *Enquiry into Plants*, the Greek philosopher Theophrastus wrote about four Indian fruit trees including what must have been the banana. He recorded that the fruit was used as a food by the sages of India. Pliny "The Elder," a Roman naturalist, added to the story in his *Historia Naturalis* in 77 A.D. by saying that the tree was called *pala*, almost identical to *palan*, a Malayan word for the fruit. These accounts were enough to cause Linnaeus in the eighteenth century to apply the term *Musa sapientum*, "Fruit of the Wise Men," to the banana.[71] Yet despite the Greco-Roman knowledge of the fruit, there is no evidence that the banana made its appearance in the Eastern Mediterranean or North Africa until the Mohammedan conquest in 650 A.D. Certainly the fruit was there by 1831 when Benjamin Disraeli, the noted British Prime Minister, wrote to his sister remarking on the delicious fruits of the region. He said, "The most delicious thing in the world is a banana which is richer than a pineapple."[72]

The history of the plantain is wrapped in legend. The Linnaean name for the banana's cousin is *Musa paradisium* or "The Tree of Paradise" as it is known in the *Koran* of the seventh century A.D. In the Christian tradition it is the "Tree of Knowledge" associated with The Garden of Eden and Adam and Eve. If Europeans believed that Adam took a bite of the apple, the source of good and evil, it is most likely because European Christians knew about apples but not bananas. Early Christians thought it may have been figs or grapes, and it wasn't until the fourth or fifth century that the apple began to appear in representations of the famous tree.[73] To an Islamic resident of Sri Lanka it was a bite of the banana that caused all the trouble in Paradise. The banana plant was sometimes called "Adam's fig-tree," which means that it is also reassuring to know that when the first couple was expelled from Eden they were wearing tent-size banana leaves rather than immodest fig leaves.

From its southeast Asian origins the banana migrated in several directions including, as we have seen, across India to parts of the Middle East. The banana was probably indigenous to southern China in the Canton region. That the fruit is not mentioned in ancient Chinese literature is due to the fact that the center of ancient Chinese civilization was in the Yellow and Yangtze River valleys to the north. Yang Fu, an official of the Later Han Dynasty writing in the second century A.D., gave one of the earliest descriptions of the banana in China in *Record of Strange Things.* He named the plant *pa-chiao* or *kan-chiao* and noted that one stem bore several tens of fruit.[74]

If you wish to aid the scientific investigations of archaeobotanists and you are planning to throw away a cooking pot, don't wash it. The scientists have used infrared spectroscopy chemical analysis of gunk—actually organic residue—from excavated pieces of pottery to determine that bananas were present on Yap island in the Caroline Islands of Micronesia thousands of years ago.[75] It is evidence that as the human population spread eastward from Asia across the Pacific the banana was carried with them. In 1774 Captain James Cook found plantains cultivated in New Caledonia in Melanesia. Even on Easter Island in the southeastern extremity of Polynesia and almost two thousand miles from the nearest human habitation, bananas are cultivated in the wind-sheltered inner parts of the island.[76]

The eastward migration of Indo-Malaysian bananas seems to have skipped the island of New Guinea north of Australia. Nevertheless, the island is undoubtedly the origin of a group of bananas known as the Australimusas.[77] Unlike most bananas, the fruit of the Australimusas does not hang pendent but grows straight up.The Fehi or Fe'i is one of the few edible bananas in this group, but it must be eaten cooked.

The migration that led to today's commercial banana moved westward from southeast Asia across Africa and on to the New World. Bananas probably were first brought to east Africa by sea in the first century A.D. by the Sumatran (Indonesian) colonizers of Madagascar. It is unlikely that the banana reached equatorial Africa by land routes from the north because of the difficulty of overcoming the continent's topography and climatic problems.[78] If you have ever had a potato sprout after storage in a warm and humid kitchen, it should be no surprise that partly dried banana rhizomes and suckers can be carried great distances and will take root and grow in rich soil and a warm humid climate. The Sub-Saharan African climate is well suited for the banana, and it is among the most productive staples of tropical African agriculture.

Al-Mas'udi, an Arabic visitor to Madagascar during the tenth century A.D., found the people of eastern Africa eating bananas and coconuts. The Portuguese began their voyages of exploration of West Africa in the fifteenth century A.D. and found plantains and bananas well established there.[79] They were probably carried across equatorial Africa from east to west by the Arabs in connection with their trade in ivory and slaves. Many of the indigenous African peoples readily adopted the fruit and aided its migration.

In Rwanda and Burundi bananas became a traditional crop mainly used to brew beer. In the latter country bananas occupy twenty-five percent of the arable land. Several species of banana are the most important crop for the Chagga people living south of Mt. Kilimanjaro in Tanzania and Kenya. For the Babali tribe of the Democratic Republic of the Congo—perhaps better known in its earlier incarnation as Zaire—the banana or plantain is the most important food crop by far. Unripe bananas are boiled as a potato and made into a porridge or they are sliced and sun-dried. In northwest Cameroon each man of the Banyangi tribe owns a farm which he divides among his wives. Each wife has a field to grow bananas, yams and peanuts. The plantain, eaten boiled or roasted is the most important food.[80]

Africa was the origin of a relative of the banana which was domesticated at an uncertain time in the highlands of Ethiopia. *Ensete ventricosum* is a member of the genus *Ensete* which with *Musa* are the only two genera of the family *Musaceae*. It is known by the common name enset and is a banana-like staple of several tribes which grow it for food and fiber. The pseudostem base yields a product that is nearly pure starch and is fermented in preparation as a food.[81] A bit of a stir was created in science circles when a forty-three million year old plant fossil belonging to the genus *Ensete* was found in 1983. Ian Gordon, now a geologist, was a high school student when he made his find in Oregon in a shale formation entombed by a volcanic eruption. The slightly curved one and a half inch long and half inch wide fossil may be the oldest ever found.[82] Oregon once had a warm, humid tropical climate, but it's a long way from Ethiopia.

The Portuguese who in their day were the greatest distributors of cultivated plants, first explored the Guinea coast of west African between 1469 and 1474. From there they carried both slaves and banana plants to the Canary Islands, Maderia and the Azores where bananas did not previously exist. European slave traders found it advisable to transport slaves along with their customary foods. Slave survival and productivity rates were better when a familiar and acceptable food supply was available. It is quite likely that bananas, rice and yams originally made it to the New World for that reason.[83]

Our word "banana" is of undoubted West African origins where the Guinean name was "banema" or "banana." In the Bantu group of languages the banana is known as "banane" and "bana." The word "plantain" was found first in Spanish as "plantano." Both "banana" and "plantain" entered English in the seventeenth century by way of Spanish and Portuguese, and the two were often written about indiscriminately. Today "plantain" may be used by Spanish speaking peoples for all types of bananas.[84]

In 1496 the Spanish displaced the Portuguese in the Canary Islands, and the stage was set for the banana's move to the New World. Not only did Father de Berlanga carry the banana to Hispaniola in 1516, but after 1530 as Bishop of the See of Panama covering the entire west coast of South America, he encouraged the cultivation of bananas, lemons, oranges and other fruit.[85] It is not known what specific banana variety the pioneering priest introduced, but many varieties quickly followed and readily took root in the tropical and subtropical regions of Mesoamerica and South America. However, to this day the greatest diversity of bananas is to be found in their main center of origin, southeast Asia. The diversity of banana clones declines in sequence from Asia to Africa and then America.[86]

When one considers the yield of the banana plant, it is no wonder that it was embraced in the New World and elsewhere. The banana is one of a dozen species of cultivated plants that account for over eighty percent of the modern world's annual tonnage of all crops.[87] The expected yield from a good acre of wheat is about 1,300 pounds a year, and for Indian corn the yield is about 2,800 pounds. An astounding 18,000 pounds of bananas can be produced on one acre in one year. For the plantain the German scientist Baron von Humboldt estimated under ideal conditions a yield per acre up to forty times greater than that of the potato.[88] Total world banana production, including plantains, is not reliably recorded, but in 1990 approximately 73 million metric tons were produced in 120 countries. About 90 percent of the total production is consumed locally as a staple food. Of the total production, thirty-six percent comes from Latin America and the Caribbean; thirty-five percent from Africa; and twenty-nine percent from Asia and the Pacific.[89]

You might say that Doug Richardson's Seaside Banana Gardens with over fifty varieties of bananas represented a culmination of the two great globe circling paths of the migration of banana cultivation. La Conchita, California, representing the extremity of the westward migration of bananas from their Southeast Asian origin, is the closest point of banana cultivation on the mainland of the American landmass to the Hawaiian Islands which represents the eastward migration of the banana across the Pacific. But starting his farm and linking the banana's migratory routes wasn't exactly easy for Richardson.

As Richardson recalls it, "I made a decision that I wanted to plant bananas, and I realized we had this little unique microclimate. I did not want to venture too far from my own little front yard there. I know that that was a rather restricted kind of situation, and I began asking people at either end of town who owned land if they would lease to me. I got resoundings 'nos.'" Two years went by, and Richardson continued his landscape work. It was the early eighties, and the economy was kind of tough. But Richardson was persistent. "I had a little supply of banana plants built up in the yard. So I asked again if they would lease, and they said no." His wife told him, "You've got to get a regular job." The aspiring farmer put together a resume and got a job as the quality control person for the biggest wholesale nursery in California. The dream was put on the shelf, and he sold off some of the plant material.

The "regular job" was an unbelievable grind which required leaving home before dawn and returning after dark. "So," he said, "I kind of just started thinking about the bananas again." Then came a serendipitous moment. He noticed that the field at one end of town was being cleared as if in preparation for planting. "I just thought well they're doing something with it, and I guess it's never going to happen. Then it just sat there, and nothing happened. So I asked the manager of the ranch after about three or four months, 'Are you guys going to do anything? What happened?'" The manager responded with the information that the owners had planned to lease the land to a flower grower who decided after a soil test not to proceed. Richardson saw his opportunity and inquired if the owners would lease to him, This time they said yes.

Richardson obtained the land in March of 1985, and began planting in June. His initial plantings included about twenty banana varieties and kept growing from there. About eighteen months later the first bananas were ready for harvest. Richardson recalls the event with satisfaction. "We started harvesting bananas and going to the farmers' market. It wasn't until two or three years after that that we built the fruit stand and began selling retail."

There were plenty of nonbelievers. As Richardson recalls it, "We actually used to park our vehicles in the back of the banana field and hide out from the public because people were so curious, and they wanted to come by and stop and tell us didn't you know you couldn't grow bananas in California. We just thought it would be better if we didn't have to talk to anybody while we did our work."

Sounding like a proud parent, Richardson relates the next step. "We opened our own stand. And it was such a dramatic thing from the freeway; we were right there on the freeway. Some writers began coming by, and we got a series of articles that just kind of built upon each other. Before we knew it, we were pretty well known." Frieda's Specialties of California heard about the operation, and Richardson began to wholesale bananas to the company which sent the bananas from La Conchita to San Francisco, Texas, and even the east coast.[90]

Mail orders soon became another marketing device. Richardson offered a sampler pack of five or six varieties. For several years ninety percent of the mail order business came from one customer. An unnamed health oriented Hollywood producer and his wife sent out a large number of banana sampler packs as Christmas gifts. The healthy gifts were well received and became the couple's signature gift. Richardson received the orders in the summer and then managed the crop so as to provide the very best for the holiday season.

Just when things were going well, the unthinkable happened. On December 22, 1990 Seaside Banana Gardens experienced about an hour and a half of 32-33° temperatures. Richardson painfully recalls the situation. "That's all it took to destroy most of our crop. At the time we had all twelve acres there just laden with fruit. It was our first really heavy crop for all three fields. It caught us and wiped out the whole thing. That was pretty tough." But Richardson and the bananas bounced back.[91] His "Palapa" fruit stand at the banana field, offering varieties of bananas not usually available commercially in the United States, became an attraction for locals and tourists alike.

No one knows for sure how many kinds of bananas there are on the planet. A high side estimate of edible bananas is around five hundred varieties, and not all of the South Pacific islands or Eastern Africa bananas have been fully classified.[92] Another estimate is that there must be about three hundred distinct bananas, one hundred fifty clones and one hundred fifty easily recognized mutants. The true number probably lies somewhere in the range of two to five hundred.[93] Richardson grew about fifty varieties at his California farm. The variety that intrigued people the most and that the first one they asked about at his fruit stand was the "Ice Cream" banana. Says Richardson, "The name for obvious reasons is catchy. The other name for the banana is 'Blue Java' which is equally intriguing. Before the banana turns yellow, while it's in the green state, it turns this incredible silvery blue. When it is well grown, it is an exceptional banana. It's definitely an exciting banana. It's a tough one to grow in our climate. It does better in a warmer climate. We have grown some really good 'Ice Cream' bananas."[94]

You probably won't find the "Ice Cream" banana in your local supermarket, and Richardson's bananas were different in other ways from the main commercial varieties. His bananas took longer to grow and ripened longer on the plant. The result was denser bananas with a lower water content and a sweeter more flavorful taste. Another major difference was that all of Richardson's bananas were grown organically. None of the herbicides, nematecides, insecticides or fungicides used on imported fruit were used at La Conchita.

Banana connoisseurs describe banana tastes with the same rapturous precision as wine enthusiasts describe a well-aged cabernet. Richardson says a slightly soft—or is it not sufficiently "aged"—"Blue Java" has a "tart pineapple-cherimoya flavor" which becomes increasingly "rich, almost liquor-like" as it ripens further. Speaking of the same banana, William O. Lessard, a Florida based banana expert, says the "Ice Cream" is sweet but has just enough acidity to offset the sweetness and make an interesting taste.

Richardson's two best sellers were the "Brazilian," the banana that propelled him into the business, and the "Mysore." The "Brazilian" when fully ripe has an intense sweetness balanced by a pleasant tartness. It grows to a height of twenty feet and is one of the few bananas that can be allowed to ripen on the plant without fear of the fruit splitting. Although not recommended, if you give the plant a shake when the fruit is ripe, you will be bombarded by banana bombs. The "Mysore" is a short fat banana sometimes called a "lady finger" banana, a generic term used to describe short fat bananas in many parts of the world. The "Mysore" is a native of India where it is about seventy percent of the crop. Richardson describes the taste of the orange-fleshed banana as having rich strawberry-citrus overtones. Lessard believes that sooner or later the commercial growers will offer this adaptable, hearty, disease-resistant banana.[95]

Some bananas fall into a category known as "boutique bananas" which means they are not readily available and cost more than the local supermarket variety. The "Manzano" is a favorite dessert banana around the world and one grown on the Richardson farm. It is a short plump fruit with flesh noticeably drier and firmer than other bananas. If eaten before it is fully ripe, it has an astringent flavor, but when fully ripe it has a pleasant apple tartness which accounts for its other name as the Apple banana. When sautéed, it develops its own rich pineapple tasting sauce.[96]

A boutique banana that you might find in the supermarket is the "Jamaican Red" also called the "Cuban Red" or "Hawaiian

Red." Lessard prefers the Costa Rican name "Macaboo." Richardson says this is true nobility among banana cultivars. Among ancient Hawaiian royalty this was the preferred banana. For commoners they were taboo. In the Sanitaria religion still practiced in the Caribbean, the "Macaboo" was a favorite fruit offered to the saints. The skin is an almost purple dark red although the flesh is a typical banana color. In the supermarket you might find this variety with a sticker which tells you it is ripe when the skin looks like a tropical sunset. Lessard says the fruit is sweet and moist with a buttery consistency.[97]

The boutique is loaded with other varieties. From the Philippines comes the "Cardaba." This ten-incher can be eaten raw as a sweet dessert or cooked as a starchy vegetable. The flesh is a pale salmon-orange. The "Ae Ae" or "Koae" from Hawaii is a beautiful plant with variegated white, grey and green fruit and leaves. As the fruit ripens the white and green strips turn a dark and light yellow. The pulp is light orange. It is a very colorful plant that thrives in a cloudy and moist environment.[98]

Hawaii is definitely banana territory. Along with those already mentioned are the "Iholene" group and the "Hawaiian Black Banana" also known as the "Ele Ele." The "Iholene" group produces a fast maturing light yellow fruit with a dark orange pulp with a very exotic flavor. The "Hawaiian Black Banana" has a dark almost black pseudostem, but the fruit is the usual green ripening to yellow. It too is so delicious that it was taboo to all but Hawaiian royalty, but the commoners planted them in the jungle anyway.[99]

The banana world has its share of oddballs. The "Rhino Horn" from East Africa resembles the shape of the animal's horn. Each fruit can grow to two feet long and can weigh as much as three pounds each—enough for an entire family. The "Rhino Horn" is virtually identical to the "Moongil" grown in Jamaica. The "French Horn" does not resemble the instrument of the same name. It is also known as the "Dominica" and is a plantain and a food staple of the American tropics. W. O. Lessard has given the name "Golden Aromatic" to a banana from Southern China. In China it is known as "Go San Heong" which means "you can smell it from over the next mountain." It doesn't stink; it's just that it has a strong pleasant fragrance.[100] While all of these banana varieties are significant in their own way, they are not the bananas of international commerce.

From 1870 until the 1950s the great banana of commerce was the "Gros Michel," or the "Bluefield" as it was sometimes known in its early days. The "Gros Michel" looks much like the typical supermarket banana of today. It produces a large and compact stem of bananas that is easy to ship, but it has several flaws. It is a tall plant—twenty feet—which makes it susceptible to blowdowns in high winds, and it tends to snap under the weight of its large heads of fruit. Its fatal flaw is its susceptibility to Panama Disease Race One, a fungus called Fusarium oxysporum that will cause the entire plant to wilt. The disease is uncontrollable in a susceptible variety. During the 1950s the United Fruit Company, the largest banana company at the time, estimated annual lost production to the disease at around $18 million annually. Entire plantations of "Gros Michel" were wiped out. It is one of the best examples of the pathological perils of monoclone culture in agricultural history.[101]

Since the commercial demise of the "Gros Michel" around 1960, the Cavendish group of cultivars has become the commercial staple of the industry. The Cavendish group includes the "Valery," also known as the "Robusta" or "the strong one," the "Grand Nain" and the "Williams." There are few differences among them. The members of the Cavendish group are resistant to Panama Disease Races One and Two, but they are susceptible to Panama Disease Race Four which has not appeared in the western hemisphere. They are also less likely to experience blowdowns because the plants are shorter than the "Gros Michel."[102] These qualities prompted the United Fruit Company to switch its Central American plantations from the "Gros Michel" to the "Valery" around 1960.[103]

Like its commercial predecessor the "Gros Michel," the "Cavendish" may be in danger. It has fallen prey to new strains of Panama Disease in Asia, and Black Sigatoka, a fungus that kills banana plant leaves, appears to be developing a tolerance to the fungicide applied to banana fields. Fields that used to require spraying ten to fifteen times a year now require fungicide applications up to forty times a year, an increased cost of production that drives smaller farmers out of business.[104]

The search is on for a commercial replacement. Dr. Philip Rowe and his colleagues at the Honduran Foundation for Agricultural Research (Fundacion Hondurena de Investigacion Agricola—FHIA) have bred the "Goldfinger" banana, also known as FHIA-01, which is genetically resistant to Black Sigatoka and has a high resistance to nematodes. But there is a problem. The "Goldfinger" has a tarter flavor than does the "Cavendish." Will the consumer accept a shorter, fatter and less-sweet banana? Doug Richardson says, "I'm very keen on it. I think it can be excellent." But he adds a note of caution. "Virtually the only criticism I have heard of the banana comes from people like Chiquita. And they say that they think the banana is too tart, too acid, it's not sweet enough, and the American public is not going to go for it." One of the parents of the "Goldfinger" is the "Dwarf Brazilian" which is an offspring of the "Brazilian" about which Richardson observes, "It is higher in acid. It's much fruitier; it's a much richer banana. Based on my experiences of eleven years of selling bananas, people prefer it. They prefer it far and away. I think it's a situation where the larger corporation, the larger company with the incredible investment, likes the status quo. They are reluctant to accept something that is new and different."[105] Any change in commercial banana variety would likely require changes in packaging, temperature control and shipping procedures—an expensive proposition.

Developing the "Goldfinger" did not require agent "007," but Dr. Phillip Rowe has been in bananas a long time. A one-time watermelon breeder, Rowe joined United Fruit in 1968 in its banana breeding program in Honduras. When United Fruit turned the operation over to FHIA in 1984, Rowe stayed on. Finding a "Cavendish" replacement was an arduous task. None of the bananas we eat today are man-made hybrids. They are all clones of bananas that were domesticated from wild ancestors over many centuries . To develop the "Goldfinger" it was necessary to find a suitable seed-producing variety from Southeast Asia that could then be crossed with a commercial variety. Rowe and his colleagues spent four years pollinating bananas—an estimated ten thousand bunches—by hand. Thousands of bunches of bananas were peeled and mashed to yield just thirty seeds. One of them produced a plant with disease resistance. The pollen of this "father" was mated with a Brazilian banana "mother," an apple-flavored "Dwarf Prata" clone, to produce the Goldfinger. Over several years the plant was successfully tested in several different soil and climate locations around the world. But it seems that the acidic taste makes the Goldfinger still a work in progress. A Dole spokesman found the banana too acidic and too starchy to be an acceptable dessert banana. The search for "Goldfinger II" goes on.[106]

Dr. Rowe and his colleague, Dr. Franklin Rosales, have also developed FHIA-02, a hybrid Cavendish highly resistant to Black Sigatoka, and FHIA-03 a cooking banana that is proving more vigorous than conventional plantains. These developments have produced high-yielding, disease-resistant, and tasty plantains which is very good news for seventy million Africans for whom plantains are a staple food.[107]

Meanwhile Canadian researchers—yes, Canada, the country where Moose in heat have been known to trample banana plants in snow-covered fields—have developed a real work of banana art. It is known as the "Mona Lisa," but it has a big aesthetic problem. It must turn a dark brown outside before it reaches ideal ripeness. It is not likely nor was it intended to replace the "Cavendish." The Canadian government spent $7.5 million over fifteen years to create a low-cost banana that can feed the hungry masses of the world. The "Mona Lisa," actually developed at a research facility in Honduras, is genetically immune to the Sigatoka fungus and thus does not need the expensive fungicides that small farmers cannot afford. Although designed with subsistence farming in mind, the "Mona Lisa" does have commercial possibilities. As an organically grown banana, it has had some success since 1996 in health-oriented markets including banana-loving Canada where the annual per-capita banana consumption is higher than the United States.[108] Now if they could just do something about that dark brown color.

The "Goldfinger's" taste and the "Mona Lisa's" color will probably keep them from becoming the commercial banana of the future. There are no potential new banana varieties with resistance to the major diseases and nematodes that could be the new dessert banana to replace the "Cavendish," but the Banana Improvement Project is trying to find one. The BIP includes seventeen research projects throughout the world. The project is sponsored and significantly funded by the Common Fund for Commodities, the Food and Agriculture Organization Intergovernmental Group on Bananas and the World Bank. Its goals are to make bananas a more profitable crop and to reduce excessive pesticide use.

Several new developments may force the banana export industry to consider finding new varieties. The costs of disease control, especially Black Sigatoka disease, have been rising, and fungicide-resistant strains of the disease have appeared. There is greater potential for environmental damage to waterways due to the run-off of excessive fungicides and nematicides. New strains of Panama Disease that have appeared in Asia and Africa threaten the "Cavendish." And finally, new virus diseases such as banana streak virus have been found in Africa.

In 2001 the Global Musa Genomics Consortium made up of nonprofit institutes and universities from eleven nations announced a project to sequence the banana genetic code within the next five years in order to genetically engineer resistance to diseases and pests currently threatening banana and plantain crops. Because most of the world's edible bananas have been in an evolutionary limbo for thousands of years, they are particularly vulnerable to pests and diseases. Thus they are one of the most heavily sprayed crops resulting in damage to both the ecosystem and the health of plantation workers. The consortium hopes its work will lead to a banana that will not require huge amounts of toxins to survive.

Banana will be the first edible fruit to have its genetic code revealed. Only two plants have been sequenced so far: the mustard cress Arabidopsis thaliana and rice. Unlike the companies that sequenced those two crops, the banana consortium will make the gene sequences freely available. It is hoped that such information will benefit the small farmers who grow eighty-five percent of the world's bananas, mostly for their own consumption. Genetic modification could make the bananas resistant to disease. But they won't be seeking a gene to make the banana straight because as one project leader observed, "It would take all the fun out of bananas. They'd be so boring."[109]

Since the mid-1990s most of the funding for banana improvement has come from public sector sources, but the private sector biotechnology industry has also moved into improving bananas. In 1994 DNA Plant Technology of Oakland, California and Zeneca Plant Science in Delaware announced they would joint forces to develop bananas that will ripen more slowly. The result would be fruit that is fresher and more flavorful. The companies planned to suppress the gene that produces ethylene, the chemical produced by the banana that hastens ripening. If successful, the banana can be left on the tree longer. A more mature fruit is more flavorful. Once harvested, the genetically altered banana would have a longer shelf life. Zeneca planned to improve the "Cavendish" variety while DNAP focused on other varieties not usually marketed in the United States, at least not yet. The Food and Drug Administration would have to approved the altered fruit for sale in the United States. Another biotech firm has already genetically engineered a tomato that can be picked later and last longer.[110] If tomatoes can do it, so can bananas.

While progress was made on the biotech front, all was not well back at Doug Richardson's Seaside Banana Gardens. The La Conchita microclimate was fine for bananas, but the weather in early 1995 wasn't. A violent winter storm caused a million cubic yard landslide. A wall of mud descended on the town and cut off access to the area for more than a month. The mud engulfed nine houses, covered streets, and buried Richardson's first field and his irrigation system.

More than one hundred La Conchita residents sued the La Conchita Ranch Co. which owned the hillside citrus and avocado orchard that slid onto the town. The suits claimed that overwatering of the orchard created the instability of the slope. The ranch company claimed that the event was a natural geologic process aggravated by heavy rain and the 1994 Northridge earthquake. A class action law suit was filed in the Fall of 1998 and some homes in the town still bore county warnings of a geologic hazard zone.[111]

The ranch company owned the land that Richardson's banana farm occupied, and although the banana entrepreneur estimated that the damage and lost business cost him $250,000, he did not sue. Instead he began to withhold some rent which by September 1997 amounted to about $30,000. The ranch company told him to pay up or leave by March 1, 1998.

Unable to work out a compromise, the banana plantation closed. Richardson was forced to return the land to the condition it was in when he planted his first bananas thirteen years earlier. Some of the plants were sold to wholesale nurseries, and some potted stock was hauled to a sixteen acre plot at the other end of town which was to serve as a holding area while Richardson planned a new botanical garden of tropical plants. Most of the farm's fifty thousand plants were bulldozed along with a eucalyptus tree windbreak along Highway 101.[112]

La Conchita lost a tourist attraction that on a pleasant weekend attracted hundreds to the tiny bit of the tropics by the sea. Angry letters to the editor deplored the treatment of a productive citizen and the destruction of healthy plants. But to look on the bright side, Doug Richardson is probably even now planting an ice cream banana just down the road or over the next ridge. Bananas have been migrating for thousands of years. There is no reason to think they would stop now.

CHAPTER 3

Banana Business

"We do our business entirely for profit. Whenever we see we can make a dollar, we do it."
Andrew Preston, 1913

The Caribbean coastal lowlands of Central America are prime banana territory. From Belize in the North down through Guatemala, Honduras, Nicaragua and Costa Rica to the isthmus of Panama in the South, the coastal area is hot, well over 75°F year round, and wet, over two hundred inches of rainfall each year. Not surprisingly, the natural vegetation is a humid tropical rain forest which looks like a carpet of broccoli when viewed from the air. At least it was rain forest until the banana men arrived in the late nineteenth century to plant their rhizomes and harvest the golden fruit. Like the tropical entrepreneurs, the banana was not native to Central America, and the arrival of foreign men and plants changed everything.

Honduras, the quintessential "banana republic," did not gain the label without reason. At 43,277 square miles, the country is slightly larger than the State of Virginia. Three-quarters of Honduras is mountainous with tolerable temperatures; the rest is coastal plains or swamps with stifling heat and torrential rains. Before the arrival of the banana, most of the overwhelmingly mestizo population lived in the cooler highlands avoiding the always hot and oppressive Caribbean coasts. The arrival of American banana pioneers on the north coast of Honduras in 1899 changed the focus of Honduran development.

Banana growing has been described as a pioneering, high risk agriculture which developed as an extension, psychological if not geographical, of the American frontier. Like many other Central American areas, the northern Honduran coast at the end of the eighteenth century was a "lawless, pestilence-ridden and almost uninhabitable territory." In the low-land tropics disease and sickness were the norm while good health was the exception.[113] Like most Central Americans who valued their lives, native Hondurans avoided the North coast. It was in this territory that the banana entrepreneurs arrived in full force in the first decades of the twentieth century.

By 1914 sixty percent of direct foreign investment in Central America came from the United States, and it was almost under the total control of two banana companies. The banana company investments included control over land, railroads, port facilities, utilities and communications systems. Because banana operations were often in isolated enclaves, the companies provided for their own general stores, banks, hospitals, clubs, and breweries.[114] The United Fruit Company was such an operation. Between 1910 and 1930 the company got control or leased nearly 400,000 acres of Honduran land. The town of Tela on the north coast was the leading banana port in the Caribbean, and the company-built Tela Railroad was United Fruit's chief Honduran subsidiary.[115]

The Tela Division was the first of United's divisions in Honduras. In 1933 it was producing seven and one-half million stems of bananas every year, the result of precise planning, organization and execution. First came the agricultural chemists to find the best soil. Then came the government concession. The company received some land as a grant; the rest they bought or leased—200,000 acres total. The company constructed a camp on the coast which became a port town with a pier where ships arrived with lumber, rails, machinery, railroad rolling stock and all the other necessary accouterments of industrial civilization. The railroad penetrated the jungle and wound its way inland along the Ulua river. With it came drainage systems, water systems and telephones. Shacks and then farm houses were built along the tracks. Surveyors went into the jungle, dislodged the alligators and laid out precise one thousand acre farms. Stetson-hatted overseers with .38's supervised barefoot laborers, known locally as mozos, who slashed the underbrush with their machetes. Rattlesnakes and fever exacted a toll, but the banana root stocks brought from Guatemala and Jamaica were planted three hundred to an acre.[116]

Young men from the States, preferably graduates of agricultural and engineering schools, lived three or four to a house along the railroad. Most of them were timekeepers. The job, at $60 a month to start, was to see that the mozos, from everywhere except the States, worked and then to pay them for what they did. As a frontier life, it was often dull and sometimes dangerous on the railroad line. The young timekeepers learned to curse fluently in Spanish if for no other reason that to break the deadly monotony. On week nights they read old magazines before falling asleep. Pay days were different. The company sent motor cars up and down the line to take its men into town; staying on the farm was not safe. In town the men could flirt with the native women or visit the company headquarters building where they could drink whiskey and shoot craps all night. Out along the track the mozos let off steam under the influence of guaro, a potent local brandy distilled from sugar cane. During the night they would often shoot and slash one another, and the company men returning the next morning often saw a circle of zopilotes enjoying breakfast in typical buzzard fashion.[117]

In his 1950 novel *Green Prison* Ramon Amaya Amador, a young banana worker for almost a decade, described the working conditions of the plantation.

> Amid all the confusion of workers and banana trees, sun and affliction, sweat and machines, creeks and malaria, cried out the conceited voices of the foreman, the whistling of the supervisors, and the arrogant cocky gringo slang. That's the way it was all day long. The exhausting work of the peasants stretched out until dusk when, their legs weary, they quit the green prison of the banana fields only to face the joyless, soulless barracks they had to live in.[118]

We are not talking gold mining here; this was the relatively humble banana that can be purchased today in any supermarket for maybe fifty cents a pound. Why then did large American corporations invest millions of dollars in foreign lands, and why would men suffer the conditions of an inhospitable environment to cultivate a fruit that perhaps not one in ten thousand Americans had seen or tasted before 1870? There is a story here that rivals anything the American "gold rushes" can offer in colorful characters, improbable situations and swashbuckling adventures.

No one knows with certainty when the first banana arrived on United States shores. It is claimed that an unnamed merchantman delivered the first shipment, probably from Panama, to Salem, Massachusetts in 1690 "where the Puritans boiled the bananas with pork to complement a boiled New England dinner. The citizens reported with great indignation that the bananas tasted like soap."[119] It is likely that vessels in the Caribbean trade occasionally brought bunches of bananas to colonial ports before the Revolutionary War, but if Ben Franklin ever enjoyed the tropical fruit, he never reported it. A more authentic report indicates that the schooner *Reynard* brought thirty bunches from Cuba to New York in 1804.[120] Through the first half of the nineteenth century small quantities of bananas were imported on a haphazard basis, and given the vagaries of sailing vessels, winds, and weather, it was likely that Caribbean bananas would arrive in North American ports in an over-ripe condition. By the 1870s imports to the United States were counted by the thousands of stems.[121]

It might be considered that it was the "Machine," interpreted in its broadest sense, that made the banana an American staple. In the first place it required the speed of steam powered vessels to get the bananas from tropical regions to North American ports while the fruit remained in edible condition. Secondly, until the end of the nineteenth century there did not exist an industrial enterprise of sufficient size and capital to undertake the huge task of producing, transporting and distributing bananas. Such organizations began to emerge after the American Civil War, but until that time, "bananas might as well have been solely a product of Mars as far as the people of the temperate zones were concerned."[122]

Carl August Franc, a German-born steward working on a steam mail ship sailing between New York and Colon, Panama (then part of Colombia), transported a few stems of bananas to New York in 1864. He made a one hundred percent net profit on the transaction, saw his opportunity, quit his job, and moved to Panama to supervise the clearing and planting of the fruit. By 1870 Franc and his brothers entered the business on a regular basis. Using three steamships, they brought three thousand stems from Panama to New York in May of 1870.[123]

Also after the Civil War, a Captain George Busch loaded his schooner with five hundred stems in Jamaica and fourteen days later sold them for a profit in Boston. Within a few years he set himself up as an agent for several American fruit firms. In 1869 he sent seven cargoes of bananas and coconuts to the United States.[124]

While Captain Busch may have initiated the Jamaican banana trade, it was a native New Englander with Jamaican bananas who launched what became the "Mother" of all banana companies, the United Fruit Company, the corporate ancestor of Chiquita Bananas. Lorenzo Dow Baker, born in 1840 in Wellfleet, Massachusetts, was the youngest of eight children of a "dirt poor" fisherman. His mother died when he was six, and at age ten he was apprenticed as a cabin boy and cook's helper to a fishing captain sailing to the Newfoundland banks. Baker attended school only in the winter when the fleet was inactive. At age fifteen he was the cook on a fishing schooner, and by age twenty he was a full fledged skipper.[125]

Baker was described as a man with large gray eyes, oversized ears and a distinctly large nose. He had a squarish bewiskered jaw and "thin lips [that] were set in a sort of permanent beginning of a smile."[126] It is reported that Baker spoke loudly in a deep and vibrant bellow that grew deeper and louder as he grew older. He prayed even more loudly. As a Wellfleet Methodist "Bible-back," Baker did not fish on Sundays. As the namesake of Lorenzo Dow, an itinerant American preacher of the Methodist faith, how could he do otherwise? For the first vessel of his own, Baker purchased an organ and got a crew member to play it. He read the Bible daily and led the crew in prayers and hymns on Sunday evenings.[127]

Lorenzo Dow Baker who brought his first shipment of bananas to Boston in 1870 (Courtesy of Library of Congress)

Baker was an "incorrigible individualist" recruiting his own cargoes and hiring his own crew, a regular "hands-on operator" and just the sort of man to take risks and seize opportunities. That is exactly what a thirty-year old Baker did while sailing his own overage fishing schooner, *Telegraph*, in 1870. Baker accepted a commission to transport a number of gold miners and their equipment to a point on the Orinoco River in Venezuela. His task completed, Baker headed toward home, but with a leaking boat, he was forced to put into Port Antonio in Jamaica for repairs. While waiting for the recaulking to be completed, he looked for a cargo that would improve the profit from his voyage. Opportunity presented itself in the form of 160 stems of green bananas offered at a price of one shilling per stem. Assured that they would not ripen for fourteen days, he made the purchase. He arrived in Jersey City, New Jersey eleven days later with the fruit ripening on deck. One dealer rejected the fruit saying that it was best for monkeys and that he was not a zoo keeper, but Baker earned a profit of two dollars per stem by selling the fruit to an Italian immigrant, Signor A. Petronelli, who in his native land had sold bananas imported from the Canary Islands. After this favorable experience Baker again sailed to Port Antonio and in May 1871 arrived in Boston with a large cargo of coconuts and bananas from which he earned a profit of $1152. By now Baker had taken a personal liking to what he described as the "dad blamed silly fruit."[128]

Baker's success encouraged him to repeat the voyage in July. On that occasion Baker met Andrew Woodbury Preston, the representative of the Boston commission produce firm of Seaverns and Company. In 1866 Preston, at age eighteen and fresh from a farm in Massachusetts, went to work for the produce firm in Boston where he met fruit schooners to buy pineapples, citrus fruits and coconuts. Preston was an aggressive salesman and didn't miss a trick. He sold to wholesale merchants and wasn't too proud to peddle a pushcart of fruit. According to Preston, when he met Baker and his bananas at ship side, "I saw 'em, I bought 'em, and I sold 'em." Thus began a relationship between the two men and the banana business that was to last a lifetime.[129]

Not every trip was a success for Baker; if weather delayed his ship, rotten fruit had to join the famous tea in Boston Harbor. It was a fourteen to seventeen day trip from Jamaica by sailing schooner. If two of every three cargoes could be gotten to Boston in time, a substantial net profit resulted. Bananas had not yet created a sensation in the United States, and the trade remained a seasonal one. In 1876 in order to inject more capital into the business, Baker established a partnership, the Standard Steam Navigation Company, with a long-time friend, Captain Jesse H. Freeman. To gain better control over the banana supply, Baker moved his entire family to Jamaica in 1877 and established a two thousand acre plantation. Preston remained in Boston to market the bananas.[130]

In the early 1880s the partnership acquired two steamships designed by Baker himself. With capacities of twelve thousand and twenty thousand stems these vessels proved so satisfactory that Baker said that they "kept Andy Preston so busy selling bananas that his pants never got shiny, but he wore out a pair of shoes every two weeks."[131] With increasing competition from the Franc brothers who had moved their operations to Jamaica, the partnership was enlarged in 1885 by the creation of the Boston Fruit Company. Captains Baker and Freeman with Preston and eight others subscribed a total of $15,000 with just a verbal agreement. With Preston in the Boston office, Freeman in charge of transport, and Baker overseeing the production in Jamaica, the company could meet the standard of Baker's motto: "Quality, Quantity and Dispatch." [132]

In March 1890 after just five years, the Boston Fruit company's original investment of $15,000 had grown to $531,000, a capital gain of thirty-six hundred percent. While other importers lost money, Boston Fruit was remarkably successful. It was decided to incorporate the enterprise as the Boston Fruit Company in the State of Massachusetts. Baker became president but remained in Jamaica while Preston became Managing Director of the Boston Division. It was Preston who decided many policy issues. It was his belief that the key to success at that particular time was in marketing not production.[133] In 1899 there were one hundred-fourteen companies importing bananas into the United States. Only twenty-two operated on any sizable scale.[134] It was a kind of free-for-all among many undercapitalized companies. There was much

The steam schooner Jesse H. Freeman, shown here at Long Wharf in Boston, was built in 1883 and used to bring ten to twelve thousand stems of bananas per trip from Jamaica. (Courtesy of Library of Congress, LC-USZ62-105210)

waste and inefficiency in shipping and distribution. Ninety-five percent of the bananas were consumed in the port cities with few arrangements to move bananas inland. In this environment the Boston Fruit Company began to widen its market into Baltimore and Philadelphia with the addition of new and larger steamships, a fleet which Preston named the "Great White Fleet." Preston also organized the Fruit Dispatch Company to sell the fruit and send rail freight cars into the interior. Newly perfected refrigerator cars helped.[135] The company and the industry were on the verge of rapid growth when the decade of the 1890s opened.

United States banana imports in 1890 totaled 12,582,000 stems, three and one-half million more than in 1889. Three million stems entered New York, two-thirds of them from Jamaica. Boston imported 1.6 million, about one half by the Boston Fruit Company, and Philadelphia brought in 1.5 million. In Baltimore, where Boston Fruit had just entered the market, 629,000 stems were imported in 1890. But the largest banana port by far, with almost 3.7 million stems, was New Orleans, the closest major port to the Central American growing regions.[136]

Being in bananas in New Orleans was tough. Competition was fierce, and many unscrupulous shippers and dishonest brokers held sway. Although the city enjoyed a reputation for tolerance, New Orleans was also for a time the focal point of prejudice against people connected to bananas. In close proximity to tropical areas with reputations for breeding yellow fever, the government quarantine inspectors at New Orleans were vigilant. There is a story of a Captain "Slumpy Gus" approaching quarantine inspection with a schooner loaded with dead-ripe bananas. One man aboard was sick, probably with yellow fever. Knowing that the quarantine officers would hold the ship for "inspection," perhaps for several days or a week, the Captain knew the bananas would rot. The old skipper threw the sick man overboard, sailed through quarantine, and sold his cargo.[137]

In mid-July 1892 the outbreak of yellow fever in Honduras struck terror in the citizens of United States Gulf port cities. The public blamed Italian crewmen moving back and forth from the tropics as the carriers of the disease. Some thought that Italians had a special capacity for carrying diseases even though most fruit boat crewmen were Norwegians in that time period.[138] Alabama prohibited the entry of "all bananas" until the first frost. The fear even reached as far inland as Mounds, Indiana where the citizens asked the State Board of Health to prohibit the passage of banana trains—about one hundred cars a day—because they were accompanied by "Italian messengers" who looked after the perishable cargo.[139]

It was to New Orleans that another banana pioneer and a founder of the United Fruit Company shipped his first bananas in 1872. Minor Cooper Keith had not intended to enter the banana business; his initial interest was building railroads. Born in Brooklyn, New York in 1848 he received a grade school education before taking his first job at $3 a week in a Broadway men's clothing store. In 1869, after a brief stint in his father's lumber business, he moved to Padre Island, Texas where he went into the cattle and hog business with an eventual herd of four thousand cattle and two thousand hogs. But his true direction in life was determined when he was invited by his brother, Henry Meiggs Keith, to Costa Rica in 1871 to help build a railroad. Henry had received a government contract to link Costa Rica from Caribbean Sea to Pacific Ocean by rail.[140] The government was seeking a way to get coffee produced in the highlands to the Caribbean coast and thence to North American and European markets. In 1871 there was not so much as a road to the Caribbean coast.[141] The task proved to be arduous.

In Coast Rica in 1871 the best access to the country and to the capital, San Jose, in the central highlands was from the Pacific coast port of Puntarenas. There was a saying about the trip from San Jose east to the Caribbean coast. A man who once made the trip to the Caribbean coast was a hero, but a man who made the trip a second time was a fool. Keith, the hero, and the first twenty-five railroad laborers, each carrying seventy pounds of provisions and tools, made the eighty-five mile trip to Puerto Limon on the coast in twelve days. Keith was assigned by his brother to organize a commercial house in Puerto Limon. He found that "not one pound of fresh beef was to be had, not a single fresh vegetable, not an ounce of ice to combat the satanic heat. All was jungle teeming with snakes, scorpions, monkeys, and mosquitoes."[142] He forgot to mention alligators.

Recruiting a labor force to push back the rain forest and the Indians was a major task. It was almost impossible to get Costa Ricans from the central highlands to work in the eastern jungles. They knew better. Keith went to New Orleans where he advertised. "Wanted for steady work. Young men, preferably single, with a liking for travel, $1 per day; food, lodging and ship fare free. Call Mr. ___." Mr.___ was a member of the police department who was desirous of exporting some tough characters. It has been suggested that the seven hundred men Keith recruited may have been victims of the convict-leasing system in the post-Civil War South. What is known is that the recruits were less than enthusiastic. Keith had to deal with the desertion of twenty to thirty recruits at a stop in Havana, and he had to put down a mutiny off the coast of Belize. Fortunately, Keith had the foresight to supply only himself and several foremen with revolvers when they left New Orleans.[143] The would-be mutineers must have had a premonition of things to come. It rained 254 inches in Puerto Limon in 1872 . The "men shivered, burned, moaned and vomited" as they suffered from malaria and yellow fever. Whiskey was the only cure. By the time a shipment of rails and ties arrived, nearly half the workers were actually dead, and the others just felt that way.[144]

During 1872 and 1873 shiploads of new workers replenished the labor supply. The first ship from Jamaica arrived in December 1872 with 123 workers for the railroad. The year 1873 saw the arrival of 653 Chinese, 130 workers from New Orleans and 894 more Jamaicans in the hell hole known as Limon. The 1873 arrivals also included two hundred southern Italians ranging in age from sixteen to sixty. By 1874 there were twenty-five hundred men at work including five hundred Chinese and one thousand Jamaicans. They were assigned to racially segregated construction camps along the line.[145]

Maintaining an adequate labor force was a problem that continued well into the next decade. Seven hundred sixty-two "strong and vigorous" Italians were recruited in 1887 with more arriving later for a total of fifteen hundred. An additional one thousand workers were added later when the deLesseps canal venture in Panama failed, and by March 1889 there were forty-two hundred men at work on the line.[146] The human death toll was frightening. At least four thousand men died from malaria, yellow fever, dysentery and other perils in constructing the first twenty miles from Limon.[147] Legend says there is a dead man under every railroad tie. It is generally conceded that without the Jamaicans and other West Indians, who seemed to be less susceptible to the health hazards of the place, the railroad would never have been completed.[148]

Minor Keith has been described as "an apple-headed little man with the eyes of a fanatic."[149] He had to be a fanatic to survive not only the labor problems but the financial obstacles as well. Construction stopped in November 1873 when Costa Rica could not get a loan in Europe. From that point until the final completion of the line in 1890, Keith was constantly running off to London to negotiate new loans, juggling figures and negotiating new contracts or extensions with the government in San Jose. Dealing with the government was made somewhat easier by Keith's marriage in 1883 to Christine Castro Fernandez, the daughter of a former President of the country.[150] Keith quickly learned that there was at least one thing that could ease his financial problems. Bananas.

Keith was the first to export bananas from Costa Rica. The rich soil, ample moisture and warmth of the Costa Rican coastal plain provide ideal banana growing conditions, a fact not lost on Keith. Almost from his first days in Puerto Limon, Keith had obtained a supply of rhizomes from Carl Franc in Panama. They were the commercially desirable Gros Michel variety. As the railroad made its way inland from the coast, so did Keith's banana planting. The bananas were freight for the railroad and deck cargo for steamships. Keith entered into a joint account shipping arrangement with Carl Franc in January 1873. Franc shipped his bananas on Keith's ships to New Orleans. Keith paid for the transport, and they split the profits. The market in the mid-1870's in New Orleans was so good that Keith expanded his holdings with a banana plantation of fifteen thousand acres in Colombia and two others of ten thousand acres in Panama (at the time still part of Colombia).[151] By 1878 Keith was chartering additional steamers for the run to New Orleans.

When Keith started growing bananas in Costa Rica, there was extensive uncultivated land in the Limon region which Keith was able to obtain from the government at "dirt cheap" prices. He established banana plantations that were usually worked by Jamaican Blacks with white managers. By 1889 Costa Rica exported 991,000 stems, all by Keith. Others did enter the business, but since Keith controlled the railroad, he also "called the tune" with regard to shipping rates and all of the exporting activities. In 1891 Keith organized the Keith Tropical Trading and Transport Company to ship bananas from Costa Rica, Nicaragua, Panama and

Colombia. In 1896 Keith began to send cargoes of fruit on liners to the English port of Liverpool although many bananas spoiled because of the distance and inadequate refrigeration.[152] Nevertheless, it was clear that there was money in bananas, and by 1899 conditions were ripe for a major consolidation in the industry.

Opportunity presented itself in the form of failure. Keith's United States marketing agent, the New York firm of Hoadley and Company, failed leaving Keith $1.5 million in debt; fortunately, the Costa Rican government came to his aid. But the real effect was to force Keith to consider an arrangement with another banana pioneer, Andrew Preston of the Boston Fruit Company. Keith's operation grew bananas in Central America and marketed most of them through New Orleans with some sales through New York. United Fruit got its bananas from the Caribbean islands, chiefly Jamaica, and sold them in the northeastern U. S. Thus an amalgamation of the two complementary companies seemed like a marriage made in heaven, or wherever such business liaisons are arranged. A merged company would see less disruption from crop failures or hurricane blowdowns in the producing areas. Another advantage would be to facilitate expansion and to gain potential efficiencies in the distribution and sale of the bananas.[153] John D. Rockefeller was making the same type of argument for oil at about the same time. Consolidation was the magic word among the business moguls of 1899.

Mr. Keith visited Mr. Preston in Boston in March, 1899, and the result of their negotiations was the United Fruit Company incorporated on March 30, 1899 in the State of New Jersey, a state more liberal with corporations than most. A representative of the Armour company, which at the time was looking for a way to control banana distribution for its eleven thousand refrigerator cars, was "watchfully waiting" to see it Mr. Keith's fiscal situation might give Armour an opening. The Armour man later wrote "We simply underestimated the resourcefulness of the indomitable, unwhippable, lovable Minor C. Keith."[154] Andrew Preston became the first president of United Fruit; Minor C. Keith took the post of first vice-president. and Captain Lorenzo Baker became a director.

The new corporation had an authorized capital stock of $20 million. The Boston Fruit Company and Keith's assets were exchanged for new shares worth about $9.2 million leaving the balance of the stock for public subscription. The new company owned 112 miles of railroad and 212,394 acres of land including 61,263 acres actually in production.[155] The company employed fifteen thousand citizens of the tropics. It operated a fleet of eleven steamships and chartered twenty to thirty more.[156] After the merger United Fruit was the largest enterprise operating in the world banana trade. But the company's appetite for bananas was not yet fully satisfied.

Andrew Woodbury Preston one of the founders of the Boston Fruit Company in 1885 and the United Fruit Company in 1899. (Courtesy of Library of Congress, LC-USZ62-105205)

Within ten months of incorporation United Fruit went on to gain control of eight of eighteen firms importing bananas to the United States. Among this group was the New Orleans independent firm of Salvatore Oteri and Michael Macheea. Oteri, a Sicilian immigrant, along with his brother Joseph, was trading in bananas since 1864 and in 1878 was credited with using the first steamship to transport bananas to New Orleans.[157]

Another firm absorbed was the Bluefields Steamship Company owned by the colorful and affable Southerner, Jacob "Jake" Weinberger. Since 1890, Weinberger, known as the "Parrot King," had been shipping bananas, coconuts, parrots and macaws from eastern Nicaragua.[158] The net result of these activities was that in the period 1900 to 1910 United Fruit's average yearly business was well over three quarters of all the stems of bananas imported by the North American and European markets combined.[159] "From the moment the green bananas were whacked from the trees until they were unloaded in the States, UFCO [United Fruit Company] reigned."[160]

Sometimes in the corporate struggle for "survival of the fittest," as in the law of the jungle, the prey escapes the clutches of the predator. Such was the case with the firm of Vaccaro Brothers and Company of Louisiana. In 1906 United Fruit paid $170,000 to acquire a half-interest in the firm.[161] About the same time the United States Department of Justice reported that United had acquired a controlling interest in fifteen of the other eighteen firms engaged in the banana business.[162] In 1908 under close congressional scrutiny for possible antitrust violations, United disgorged its interest. Joseph, Luca, and Felix, the brothers Vaccaro, and Salvador D'Antoni had formed a partnership in 1897 while managing a truck farm and selling produce in New Orleans. All the partners were Sicilian immigrants except Felix, who was born in the United States. They bought a plantation down river from New Orleans where they grew oranges. No sooner were they in operation than they were wiped out in January 1899 by the Deep South's severest winter weather on record to that time. Joseph made a decision that the partners would enter the banana trade.[163]

Sometime in 1899 D'Antoni began to purchase tropical fruit in Honduras and shipped it to Joseph Vaccaro in New Orleans. The cargoes included coconuts, oranges, plantains and bananas which were still a risky cargo on a sailing ship. The firm chartered a steamship which carried six thousand stems on its first voyage from La Ceiba, Honduras to New Orleans in February 1900. Between 1900 and 1902 the firm leased nine more banana carriers to carry fruit to New Orleans and Mobile. It was still a small operation compared to United Fruit's thirty-six steamers in its "Southern Fleet" alone. Even the addition of steamships by any company could not overcome the continuing problem of port quarantine for ships coming from tropical regions where yellow fever was present. Fifteen day fumigation periods killed not only the threat of disease but the hope of profit as cargoes of rotten bananas were dumped in the Mississippi River.[164]

In the next two decades the Vacarro firm strengthened its operations in La Ceiba, Honduras by starting its first commissary in 1912 and the first hospital in 1913. Also in 1913 the firm started Banco Atlantida, only the third bank in Honduras, and expanded its water company. A brewery producing "Salva Vida" opened in 1916 and public electric service started in 1917. The company was growing, and in 1923 it made its first public stock offering as the Standard Fruit and Steamship Co. The Vacarro brothers held the vast majority of stock but there were at least one hundred fifty other shareholders. Standard, which in its present existence markets bananas under the Dole label, was at the time the second largest banana company in the United States.[165]

The entrepreneurial efforts of banana pioneers like Baker, Preston, Keith and the Vaccaro brothers put the yellow fruit on American tables and in lunch buckets and firmly established their corporate enterprises, but it was a Jewish Russian immigrant who lifted the industry and especially the United Fruit Company to new heights. When Samuel "Sam the Banana Man" Zemurray (né Zmuri) retired from the presidency of United Fruit in 1951, he presided over a company that imported one half of all the bananas consumed in the United States. The company in 1949 operated about fifteen hundred miles of railroads and the "Great White Fleet" of fifty-two ships. United employed eighty-three thousand people, mostly in the banana growing regions, and operated schools, hospitals, a radio-telegraph service and a Honduran newspaper.[166] Zemurray could well have been a poster boy for American capitalism.

Zemurray was born on January 18, 1877 in the Bessarabian (now Moldova) city of Kishinev and at the age of fifteen emigrated with his aunt Hattie Dinitz to Selma, Alabama. Young Sam's first job was with a decrepit peddler, a "Civil War relic," who was bartering tinware for pigs and poultry throughout the South. Sam caught the pigs for pay of $1 a week. "I could outrun any pig in Dixie," said Zemurray. Later he worked as a baker's delivery boy, a housecleaner and a lathe turner. By age eighteen he managed to save enough money to send for his entire family.[167]

Zemurray's entry into the banana business came when he was eighteen and working in Mobile, Alabama as a deck laborer. He was intrigued with the possibilities of the "ripes and turnings," bananas that while en route from the growing areas had ripened to the point that they had to be sold quickly or dumped. Zemurray bought $150 worth of the ripes and arranged to ship them by rail to Selma. Legend holds that Zemurray telegraphed ahead that he had bananas to sell at each stop. Each telegraph operator got a free bunch for spreading the word. Zemurray arrived in Selma with a $35 profit and a ripe desire to sell bananas. As he enlarged his territory the cry, "Sam the banana man is coming," preceded him. Three years later, at age twenty-one, Zemurray had $100,000 in the bank. By 1900 he was a banana jobber dealing in first-class fruit and ready for the big time.[168]

To pursue this goal Zemurray joined forces with Ashbell Hubbard, an older competitor in Mobile. The partners, aided by an infusion of cash from United Fruit, bought two rusty steamers for $60,000 from the bankrupt Thatcher Brothers Steamship Co. The Hubbard-Zemurray Steamship Company escaped United Fruit's embrace a few years later when the giant fruit company sold its interest under threat of government trust-busting. Over

Unloading fruit at Mobile, Alabama around 1890.
(Courtesy of Library of Congress, LC-USZ62-105169)

the next several years the two dilapidated tubs more than returned their initial investment. Zemurray decided that the real money lay in growing the bananas and shipping them to the United States. Demand for bananas in the United States was high, and there were plenty of bananas to be had in Central America. Zemurray headed for Honduras.[169]

Zemurray arrived in a steamy Puerto Cortes on the north coast of Honduras in 1905 and launched a career in the tropics that was to become legendary. With no extradition treaty at the time, Honduran coastal towns like Puerto Cortes, were havens for people on the run, soldiers of fortune, tropical tramps and more than a few professional ladies. Americans mingled with Chinese, Syrian and Turkish storekeepers, railroad workers from everywhere and a few descendants of the Maya. One American writer on the lam from a bank embezzlement charge, William S. Porter, better known as O. Henry, captured the essence of the place in the best selling 1904 novel *Cabbages and Kings* which ridiculed Honduras as an opéra bouffe. He wrote of Honduras as a nation of greedy rulers, compliant bureaucrats and innocent peasants where Yankee banana men could make quick and easy fortunes and wield great power. Amazingly, Honduras honored the author with a statue in the port city of Trujillo. O. Henry was pretty much on target. From the time it gained its independence in 1838 to 1900, Honduras had sixty-four presidents with an average rule of less than one year.[170] Before Zemurray could make his fortune, he had to acquire land to produce his bananas, build a railroad to transport them to the coast and make the political arrangements for it all to work. It was a good thing that he was an adventurous risk taker.[171]

Zemurray began buying bananas from independent planters and shipping through the port of Omoa, but he soon decided that he had to grow his own bananas. His first purchase in 1910 was for a five thousand acre tract along the Cuyamel River in Honduras. He paid $200,000, mostly in borrowed funds, for the land. By the end of the year he owned fifteen thousand acres and was deeply in debt to bankers in the United States. The indebtedness frightened off his partner. Undaunted, Zemurray incorporated as the Cuyamel Fruit Company which was regarded at the time as just another banana firm. But Zemurray was not just another gringo. He plunged further into the jungle in search of bananas and land. He learned to speak Spanish, albeit with a pronounced Russian accent, and in three languages he could out curse anyone in the tropics. Patróns and peasants liked his tough but fair dealing. Everyone knew that "El Viejo," freely translated in Spanish as "the boss," could be trusted. He built railroads to carry the bananas to his small fleet of steamers, and he built a sales organization in New Orleans. Visions of bananas on an breathtaking scale danced in his head.[172]

Zemurray quickly realized that for his enterprise to flourish he would need concessions from the government. In 1910 the heavy indebtedness of the Honduran government to European interests threatened Zemurray's concessions. The Honduran President, Miguel Dávila, was negotiating a loan with the American banking house of J. P. Morgan. The deal to be incorporated in a new commercial treaty between the United States and Honduras would mean that the European creditors would be paid, and the bank's agents would sit in Honduran customs houses, collect duties and authorize concessions. Zemurray went to Washington. As he explained to Secretary of State Philander C. Knox, "I was doing a small business buying fruit from independent planters, but I wanted to expand. I wanted to build railroads and raise my own fruit. The duty on railroad equipment was prohibitive—a cent a pound—and so I had to have concessions that would enable me to import that stuff duty free. If the banks were running Honduras and collecting their loans from customs duties, how far would I have gotten when I asked for a concession? I told him: 'Mr. Secretary, I'm no favorite grandson of Mr. Morgan's. Mr. Morgan never heard of me.' I just wanted to protect my little business." The Secretary's only advice was that the banana man should not try to spoil the arrangement between the Morgan bank and the Dávila regime.[173] It was the age of American "dollar diplomacy."

What could an aspiring banana tycoon do? He could change the Honduran government before the deal could be consummated. As one observer noted, "In Central America it's cheaper to buy a politician than a mule."[174] Banana magnates had been playing the Honduran political game for years. When Honduran governments got low on money or soldiers they generally got both by pressuring the banana companies. Accordingly, the banana men bribed presidents, loaned money to desperate governments and paid off local chieftains to keep warring Hondurans away from the north coast banana areas.[175] Zemurray simply maintained the tradition.

Zemurray went to New Orleans to meet with the three other principals in the forthcoming revolution: Manuel Bonilla, an ex-president of Honduras and foe of Dávila, and two American adventurers and soldiers of fortune, "General" Lee Christmas and Guy "Machine Gun" Molony. A Hollywood writer with the wildest of imaginations could not invent Lee Christmas; he was real. Leon Winfield Christmas was born in Louisiana on Washington's birthday in 1863. In 1880 he began railroading at the age of seventeen and rapidly advanced to became a locomotive engineer. On a run from New Orleans to Memphis in 1891 he ran by a signal flag and crashed into another train. He had been awake for fifty-four hours and was suffering the after-effects from a party. He lost his job, and after a color test was instituted, he could not be employed by any American railroad. He was color blind.

Christmas then turned his attention to Central America where he was to develop a reputation as a "mighty drinker and mighty lover . . . [and] mighty fighting man." He became a "military leader 'worthy of an army corps,' and a private citizen hardly above the status of tropical tramp."[176] He got an engineer's job on the banana railroad out of Puerto Cortes and became an involuntary participant in an 1897 revolution against Honduran President Policarpo Bonilla (no relation to Manuel Bonilla). The revolutionary supporters of Enrique Soto captured Christmas and his train at Laguna Trestle. He was ordered at gunpoint to run the train at full throttle across the trestle against the government troops. He apparently performed the task with such enthusiasm and successful effect that he was immediately made a captain in the Sotoistas. The revolution eventually failed, but Christmas landed on his feet in Guatemala, the neighboring country that had supported the revolution. With free liquor and unlimited women, the new "captain" was a privileged character, and the legend had begun.[177]

When the next squabble over the Honduran presidency broke out Christmas was in his glory. "Rifle under his leg, twin Luger automatics holstered on his thighs, bandoliers across his shoulders, he rode . . . [with] a puro [a vile cigar of pure tobacco] cocked in the corner of his wide, hard mouth . . . ." This time he was on the winning side as Manuel Bonilla took charge of the government. Christmas became a Brigadier General and Chief of the Federal police.[178]

In 1907 a war broke out between Honduras and Nicaragua over a land dispute. A Nicaraguan invasion gave Christmas another chance to enhance his reputation. The war did not go well. Surrounded by hostile forces, Christmas led a wild horseback charge to break out of a seemingly hopeless situation. He took an enemy shot that shattered one leg and killed his horse. With only four bullets in his Luger and pinned helplessly under the weight of the fallen horse, he was prepared to save the last shot for himself. He fired at the Nicaraguans swarming toward him and then placed the gun to his temple and pulled the trigger. Nothing. Miscount. A soldier lunged at his head with a bayonet. It was but a glancing thrust. Christmas poured forth the foulest epithets he could summon. A second thrust was turned aside by a young Nicaraguan *teniente*. Christmas unleashed a new fusillade of profanity. The officer calmly replied that the prisoner would be given a formal execution, regarded by Christmas as the ultimate indignity. By now the fallen general was surrounded by a ring of soldiers.

"Shoot and be god-damned to you," Christmas raged. "Shoot now, if you've got the guts. But do me one favor. Don't bury me."

"Don't bury you?"

"No. Don't bury me, you sons of bitches!"

"But why not?"

"Because I want the buzzards to eat me, and fly over you afterwards, and scatter white droppings on your god-damned black faces."

Christmas was sure the supreme insult would bring instant death. Instead it was met with laughter.

"You are a brave man and shall not be executed at all," replied the *teniente.* All present grinned their approval.[179] While accounts vary as to how Christmas extricated himself from this situation, the "General" ended up back in Guatemala where, as the head of President Cabrera's secret police, his word was law.

In 1910 Christmas again became involved with Manuel Bonilla in a revolution against Honduran President Dávila. The revolution failed after it started prematurely, and most of the Manuelistas were captured in Puerto Cortes. In the process Christmas met Guy Molony who was destined to be the "General's" sidekick in later adventures. Molony was a veteran of the Boer War, served for a time in the United States Army cavalry in the Philippines and saw some service in Nicaragua. He was known as a genius with machines guns. By late 1910 the careers of Bonilla, Christmas, Molony and Zemurray were on a converging course with the goal of once again changing the Honduran government.

Bonilla decided that a ship was needed to be successful the next time. The *Hornet,* a vessel built in 1890 for Henry M. Flagler and later outfitted with some armor plate for use by the United States government during the Spanish-American War, fit the bill which was conveniently and secretly paid by Zemurray. The presence of Christmas and Bonilla in New Orleans to refit the *Hornet* attracted attention in high quarters in view of the ongoing negotiations between Washington and Tegulcigalpa. The conspirators were kept under constant surveillance by not-so-secret agents of the United States government. Officials were reluctant to allow the *Hornet* to leave port until they were certain that no weapons or potential revolutionists were on board. Finally, the *Hornet* cleared for Honduras on December 20 with a load of coal, but the government agents smelled a rat. Thus, the agents watched closely the next day as the party of conspirators sallied to Madame May Evans' establishment in what was then known as "The District."[180]

"Champagne corks popped in the smoke-filled air, the piano played incessantly, and Madame Evans' female 'house guests' shrieked with laughter as the playful gentlemen pinched their sterns."[181] While a government agent shivered outside in the cold December night and watched through the frosted window, inside the commander-in-chief of the revolutionary armies led the revelry as evening turned to early morning. When the last agent went home thinking that what he had watched was simply the same ostentatious behavior he had observed before, Christmas lit a puro and turned to Bonilla. "Well, compadre," he observed, "I've heard about 'em rising from rags to riches, but this here's the first time I've ever heard tell of somebody going from the whore-house to the White House."[182]

The party headed to Bayou St. John where they boarded Zemurray's forty-foot cabin cruiser and headed down the Mississippi to the Gulf of Mexico. Two nights later, safely in international waters at a good twelve miles off the coast of Biloxi, Mississippi, the cruiser pulled along side the *Hornet.* A case of thirty rifles with three thousand rounds of ammunition and a machine gun were transferred from the smaller vessel. Seeing Manuel Bonilla shivering in the raw wind, the banana man removed his luxurious overcoat and placed it around the shoulders of the once and future President of Honduras. "I've shot the roll on you and I might as well shoot the coat too," said Zemurray as he bade farewell. The *Hornet* headed south to make a revolution.[183]

By the time the United States Navy caught up with them, the revolutionists had already taken the port of Trujillo and had moved on to the port of La Ceiba. British and American gunboat commanders arranged a neutral zone in the town and urged the Dávila government commander to surrender to Christmas and the rebels. Government troop morale was so low that they were locked in the barracks to prevent desertion. The battle, if it can be called that, consisted of Molony with six men and a Hotchkiss gun attacking along the beach followed by Christmas and the infantry. Curious bluecoats on the U.S.S. *Tacoma* and U.S.S. *Marietta* watched from off shore. The defenders fled to the neutral zone and stacked their weapons before the British and United States officers.[184] The total number of combatants probably did not exceed one hundred. Tegulcigalpa, the capitol, fell without a battle.

Only four members of the Honduran Congress voted to ratify the Davila treaty with Washington. A grateful President Bonilla granted Zemurray concessions for the next twenty-five years, and Honduras was ready for a real banana boom.

Zemurray and his Cuyamel Company set out to give United Fruit, by now known as El Pulpo, "The Octopus," all the competition it could handle and then some. Few realized it at the time, but Zemurray had an advantage. United Fruit was rigidly controlled by its desk bound Boston management that left few decisions to its managers in the banana lands. "We'll feel our way—chew fine and spit carefully," said Andy Preston, United's first president. While Preston was carefully chewing, Zemurray was huffing and puffing, and supervising, and deciding, and acting. Where others built levees to keep flood waters from farms, Zemurray allowed the waters to flood lands not in cultivation. When the water drained, bananas planted in the silt-rich soil provided double their previous yield. He risked millions on extensive irrigation, selective pruning, and propping trees to prevent the heavy fruit from bruising on the ground. The result was bananas the equal, if not superior, of anything United produced.[185]

Zemurray, "that little fellow" as some United executives called him, was a pesky competitor. On one occasion both Zemurray and El Pulpo wanted to buy a fertile five thousand acre property. Ownership of the tract was in dispute between two native claimants. While United's attorneys dallied to sort out the rival claims, Zemurray paid both claimants for the land making further legal maneuvers irrelevant. In 1928 United got some measure of revenge when it bought a huge tract of land in Guatemala. While only a portion of the tract was suitable for bananas, the water rights that came with the purchase denied some water to Cuyamel.[186] Nevertheless, Zemurray's fortune was growing. By 1929 he had banked between twelve and fifteen million dollars, had thirteen steamships carrying bananas, owned a sugar plantation and a factory. Legend has it that in 1929 he bet a friend that Cuyamel's rising stock would meet United's stock going down. In January 1929 Cuyamel sold for $63 while United was at $158. On October 18, 1929 both stocks sold for $124.[187] United decided to buy out the persistent trouble maker.

In 1930 a deal was struck; Zemurray sold all of his holdings in exchange for $31.5 million in United stock, about three hundred thousand shares. Zemurray thus became the largest stockholder prompting some cynics to remark that the "fish had swallowed the whale." Mr. Zemurray agreed not to operate any competitive business. That same year United sold more bananas than at anytime before or after, sixty-five million stems which is equal to 8.1 billion individual bananas or about sixty-six bananas for every man, woman and child in the United States at the time. At age fifty-three Zemurray retired to his estate north of New Orleans to hunt, play golf and enjoy family life.

The Great Depression and the mismanagement of United Fruit changed Zemurray's plans. In 1930 New York newspapers reported that eight jobless young women in Manhattan had survived for more than a month on a diet of five bananas a day for the eight girls. They had cut the bananas into eight equal portions meaning that each girl received about five cents worth of banana a day. A banana spokesman explained that there was more nutrition in ten cents worth of banana than ten cents worth of any other food. Despite such favorable publicity for bananas, United's bottom line was not healthy. The company's profits dropped from $44.6 million in 1920 to $6.2 million in 1932. In June 1932 the stock hit a record low of 10 1/4. With a loss of eighty-five percent of his stock value, Zemurray was concerned. After talking to friends in the industry, he concluded that the company was mismanaged. The principal stockholder went to Boston.[188]

Zemurray appeared at a July 1932 meeting of the United Fruit Company's Board of Directors in Boston where he presented his charges of mismanagement. Smiling thinly, Daniel Gould Wing, the chairman of the First National Bank of Boston, said with reference to Zemurray's Russian accent, "Unfortunately, Mr. Zemurray, I can't understand a word you say." Legend says that Zemurray dumped a stack of proxies on the table before the Board composed of Boston Brahmins and stated, "You gentlemen have been f.... up this business long enough. I'm going to straighten it out." The Board understood him and appointed the man from New Orleans "Managing Director in Charge of Operations." It was clear who was boss.[189]

In the tropics the news that Zemurray was back prompted barroom celebrations among field managers and banana cowboys. The boss headed back to Central America where he replaced men whose methods he did not like with Cuyamel veterans and, operating on the motto, "You're there, we're here," he gave them greater autonomy. He cut superfluous employees and ordered ships to sail only with full cargoes. Within two weeks of Zemurray's takeover, United stock had climbed to $26.[190]

For the next twenty years Zemurray shaped United Fruit, serving as its President from 1938-1948 and from 1949-1951. His business style was definitely different. He refused to have a private secretary and preferred to communicate by telephone. He disdained paper work, a characteristic demonstrated when a manager of the Honduras division proudly gave him a long and labored financial report of ninety pages. The key dollar amounts were summarized on the cover. Without reading the report, Zemurray ripped off the cover and stuffed it in his pocket. "Most sensible damn statement I ever saw," he succinctly stated.[191]

Zemurray also took a "no frills" approach to his personal habits. He maintained a lean physique on a tall frame with a daily routine that saw him rise at six o'clock to breakfast on warm water followed by a half hour of deep breathing and stretching. As an aid to digestion he sometimes stood on his head for fifteen minutes at a time. When he needed new ideas he retired to his New Orleans plantation where he would go without eating for a week or two. He believed that a complete fast, except for water, restored his mental energy.[192]

During his tenure a tough minded Zemurray tackled whatever problem came his way. When Black Sigatoka disease appeared in the banana fields in 1936, it could have wiped out the company in a few years. United's scientists were horrified when Zemurray ordered the spraying of five thousand acres of bananas with an experimental solution known as Bordeaux mixture. It worked, and it probably saved United.[193]

No less serious in the thirties was how to deal with the rise to power of the Nazis in Germany. In the mid-1930s United Fruit entered a barter contract with the Germans whose currency was very soft at the time. The Germans loved bananas and still do. Germany agreed to built two ships in her shipyards for United's "Great White Fleet" in exchange for bananas. When word leaked that the Germans planned to confiscate the second ship for their own use, an American skipper, at United's direction, commandeered the ship at launching and took it to France and then the United States.[194]

United Fruit's "Great White Fleet" went to America's defense in both World Wars with its ships serving as troop transports and supply ships. Many ships and crews were lost in the effort. United Fruit lost a dozen ships in the first war; seven of them were owned by Elders and Fyffes, United's British subsidiary. One of United's vessels, the *Ulua*, in multiple trips, carried a total of 728 officers

Samuel "Sam the Banana Man" Zemurray in 1951, the year he retired from the presidency of the United Fruit Company. (Courtesy Eliot Elisofon/TimePix)

and 15,344 troops from the United States or Canada to Europe during the First World War. On an early voyage returning troops after the war the *Ulua* almost experienced catastrophe. As the ship entered New York harbor, the troops rushed en masse to portside to glimpse the Statue of Liberty and almost caused the ship to heal over. The incident caused a rule adoption for every returning troopship then and in World War II. Troops were ordered to remain in equal groups on either side of the ship to prevent a recurrence. The ship gained a bit more attention at a later time when Richard M. Nixon chose the *Ulua* for his honeymoon.[195]

Some of the "Great White Fleet" ships that went to war in World War II were built under terms of the Jones-White Act of 1928 which provided for low interest, long term loans for ships built in American yards. The incentive of mail contracts was also offered in exchange for building ships with designs influenced by Navy Department considerations.[196] Even before Pearl Harbor many of the ships of the "Great White Fleet" were being painted war-time gray. United provided the War Department with 113 ships during World War II; the Axis powers sank thirty-one of them. Likewise, the ships of the Standard Fruit Company were requisitioned during the war. Standard operated nine company vessels, thirty-five ships through other agents, and twenty-six Liberty ships, including number 1000 launched in 1943.[197]

Edward L. Bernays, known as the father of public relations, told a revealing story about Zemurray and the war. Newly hired as a consultant during World War II, Bernays met with Zemurray to discuss several projects. During the discussion a secretary entered the room and discreetly placed a folded note in front of Zemurray. The executive listened attentively to Bernay's idea, "nodded thoughtfully," excused himself and turned up the piece of paper and read the message. After a moment he returned his attention to Bernays and asked him to continue. This process was repeated twice more. Each time Zemurray's response was so self-contained that Bernays thought the notes contained only the most trivial of matters. Only long after the meeting did Bernays learn from someone else that the notes, delivered within the space of only a few minutes, had informed Zemurray that German submarines had torpedoed and sunk three freighters of the "Great White Fleet."[198]

Among United's wartime losses, the *Esparta*, one of three original refrigerated ships built for the "Great White Fleet"in 1904,

The U.S.S. Tarazed *was built in 1932 for United Fruit. As the S.S. Chiriqui, it was part of the "Great White Fleet" until requisitioned by the U. S. Navy in 1941 for wartime service. It was returned to United Fruit in 1946. (Courtesy of Library of Congress, LC-USZ62-105172)*

was torpedoed off the Florida coast on April 9, 1942 with the loss of one crewman and its load of bananas. The *Heredia*, running without lights and carrying a cargo of bananas and coffee beans, was torpedoed on May 19, 1942 on its way to New Orleans. Thirty men were killed. The *Parismina* went down in the North Atlantic on November 18, 1942 with a loss of life of fifteen.[199]

Under wartime circumstances the British halted banana imports in 1940. Bananas were considered a luxury and thus dispensable. Americans had to make do with a greatly reduced supply. From 1931 to 1939 an average of fifty-five million stems of bananas had been imported to the United States. Imports dropped to twenty-seven million stems in 1942, and reached a low of twenty-five million in 1943. Americans were not able to return to normal consumption levels until the post-war year 1946 when fifty-four million stems reached the United States.[200]

United Fruit's domination of the market between 1900 and 1951, a period during which it never suffered a net loss, earned the corporation $825 million, an average of 12.5% per year return on its net worth. By 1951 the Zemurray era at United was drawing to a close. In assessing his contributions, the *Encyclopedia Judaica* stated, "During his career Zemurray was probably the most enlightened of the big U. S. businessmen operating in Latin America. He endowed clinics, housing projects, recreation facilities, and schools for the workers on his Central American plantations at a time when such a course of action was considered visionary, if not lunatic."[201] Edward Bernays took note in the early 1940s of Zemurray's vision and ability to take the long view when the public relations man asked the boss about plans for crop diversification which included the planting of mahogany trees. Bernays asked how long it would take for the trees to mature. "Sixty years," replied Zemurray.[202]

The United Fruit Company played a very large role in the effort to eradicate diseases such as yellow fever and malaria. But, good health also served the labor needs of the company and protected its white North American employees. Nutritional factors also affected the death rates but got little attention from the company.[203] A former company public relations man reflected on life on the company's plantations in Central America. "To be employed as a worker was to submit your life and your entire future. And once you had submitted, no country or ideology or system of government on earth had more power over you. That was the price."[204] Nevertheless conditions on the plantations were usually better than off company property.

Zemurray paid wages higher than the prevailing rates, but at the same time, he did his best to thwart the development of independent trade unions.[205] In a 1951 assessment of his policies, Zemurray reflected, "I feel guilty about some of the things we did . . . . We've learned that what's best for the countries we operate in is best for the company."[206]

United Fruit's Annual Report for 1950 stated that it was the company's best year ever. Production was up seven percent, prices were steady, and the "Great White Fleet" had sixty-seven ships.[207] In 1951 the company had liquid assets of $50 million and a stock price of $65 a share. The sixty-six thousand shareholders earned $4.32 per share. The company controlled one-third of world commerce in bananas and shipped thirty million stems to the United States and Canada and another five million to Europe. Of the three million acres the company controlled, 139,000 acres were planted in bananas, 102,000 acres grew sugar cane, and the rest were reserves—mostly to prevent competition. At its peak in the 1950s, United Fruit was the sole economic support for a total population of almost a half million workers and dependents.[208]

Zemurray retired from United's presidency in 1951 but remained a director. A United employee wrote, "So by the time Samuel Zemurray retired, his company and his character were almost exactly matched: both were tough, no-nonsense, quick to act." The company's product was a staple of the American diet, but troubled times lay ahead for United Fruit and the industry which is probably why Zemurray, when he saw which way the company was going after his retirement, sold his entire interest in United Fruit down to the last share.[209]

CHAPTER 4

The Immigrant Banana

"If the Americans are eating it, I'm gonna eat it. I got to find out whether you eat the peel or you eat the inside."
Sara Milanow Rovner. 1991

"I was telling my daughter, you see that? I remember that. You see the watch? I remember the watch. You see this? I remember that. She says, 'Ma, I can't believe it.' After all I'm eighty-three years old, and I says to her, I remember every little bit. I can not forget what I went through those twenty-three days not knowing if my mother and father forgot me."

Arriving at the island this time was different. Yes, the Colgate Factory clock, "the watch," was still there, but this time there was the limousine ride into the city, the crossing of the harbor by private government launch, the warm official greeting, and yes, the excitement of the appearance on CNN. Josephine Cirella's first arrival in 1923 as a ten-year-old was frightening. Herded off a ferry with her mother and two older sisters, with no explanation and no chance for the merest of good-byes, she was separated from her family and detained in the island's hospital for twenty-three days. Now, in 1994, she was somewhat of a celebrity as she answered questions about her experiences from high school students around the country in a CNN live electronic field trip.[210]

A decision Josephine Cirella made in 1992 to tell her story brought about her new celebrity status. Since 1973, over eighteen hundred other people have told their story. Her experience began in much the same fashion as that of all who have recorded their recollections of anguish and hope, pain and joy, trepidation and expectation, and often considerable embarrassment and humor.

"This is Janet Levine for the National Park Service, and I'm here today with Josephine Cirella, who came from Italy in 1923 when she was ten years old. Today is August 15th, 1992, and we're here at the Ellis Island studio."[211] For thirty years the National Park Service's Ellis Island Oral History Project has been recording the experiences of immigrants who entered the United States through Ellis Island as well as the stories of those who worked there between 1892 and 1954. Dr. Levine and her colleagues are now in a race against time to capture the "grass-roots" Ellis Island immigrant stories of fading old-timers. Levine and her former colleague, Paul Sigrist, have compiled a precious tape archive of the remembrances of the tired, the poor and the wretched that both amateur and professional historians and genealogists will find to be authentic grist for the mill of their endeavors.

Between its opening in 1892 and 1924 when quota laws greatly reduced the flow of newcomers, Ellis Island was the point of entry for sixteen million immigrants representing seventy-one percent of all immigrants to the United States during that time. When the *Giuseppi Verdi* arrived from Palermo, Sicily, carrying Josephine Messina Cirella along with her mother, Maria, her two older sisters, Maria and Angelina, and two trunks, the new arrivals had to wait on board for seven days because Ellis Island was so crowded.

> I could remember when I saw the Statue of Liberty for the first time. Somebody yelled, "The Statue of Liberty, the Statue of Liberty!" We all ran outside and everybody called somebody else on the boat. The Statue of Liberty, I remember that and it was my birthday that day. Ten years old. That's why I remember a lot. Then when we docked, we couldn't get off the boat. So my poor father got stuck coming everyday, seven days, with a tugboat. And that's how I saw my father for the first time, from all the way up on the

railing down to, he was on the tugboat. And so that's how they used to bring us the bread; they brought us the fruit. But when it came to the banana, that was the best. That was the best. Everybody was yelling at one another. Look, look at those bananas. So we didn't know what they called them. So my father yelled up, because, God bless him, he had a beautiful voice, being Italian. "They call them BA . . . NA . . . NA!" He says, "BA . . . NA . . . NA!" [212]

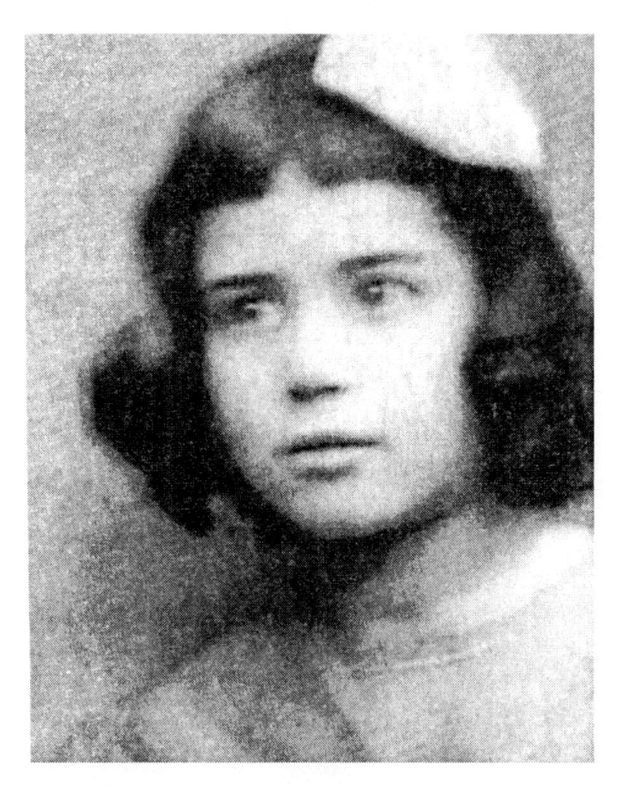

Josephine Messina Cirella when she arrived in the United States in 1923 at age ten. (Courtesy of Jo Ann Nassour)

While bananas were not unknown in the Europe of the early twentieth century, they were not part of the daily fare of most Europeans, especially those of meager means and rural backgrounds, the very same people flooding through Ellis Island doors in the early decades of the century. As a young girl from Zhitomir in Russia, Sophie Shuman Taub recalled seeing bunches of hanging bananas in big fruit stores in Kiev. "I didn't know what it was. Ooh, the rich people eating that; what is that?" At Ellis Island she saw some African-American people take bananas from a large stem of many bunches. To her wonderment, they peeled and ate them. "They're eating it; what is that?" But bananas were not the only new discovery. Taub remembers seeing big bowls of Jello. "It was shaking . . . . it shivers." She didn't eat it.[213]

The motivation for pulling up roots, the shipboard misery of the ocean crossing, and the daunting challenge of entering and adjusting to the new land are standard fare of the oral history interviews. Levine never asks leading questions about bananas, but they surface frequently as the stories of coming to America are recounted. Sometimes the first encounter with the curved fruit was before even embarking on the steerage class voyage; sometimes the initial experience was on board. More frequently, the immigrants were bedazzled by the banana after setting foot on the new soil. For some, bananas became their business. But in all cases knowing how to eat a banana was part of becoming an American. As Levine sees it, "The banana seems to become a metaphor for what was new and how they approached it."[214]

Knowledge of the banana seems to have been a sign of sophistication that early arrivals could display to friends and relatives who followed later. It was quite common for fathers and husbands to precede other family members to the United States. Josephine Cirella's father, Salvatore Messina, had arrived from San Cataldo in Sicily a full six years before his daughter. Except for a return to Italy for a stint in the army during World War One, he was an experienced hand with the "BA . . . NA . . . NA" when he brought bananas to his family on board ship in New York harbor in 1923.

Eight-year-old Angela Garofalo Basso, well acquainted with prickly pears and oranges in her native Sicily, arrived in 1920 aboard the *Dante Alighieri*.

> I remember when the boat docked and, you know,

> these small boats come along the ship, you know, selling things. And my father, they had a basket, lower the basket down with a rope, and they bought some bananas. I never forgot that. I never saw a banana in all my life. And, of course, he knew what it was. And they brought the bananas up and he peeled it and gave me one. I tasted it. Uhh, I didn't like it at all. And that's the best thing I like now. I never forgot that. That I remember. I don't know what else he bought, but I remember the banana in particular.[215]

While the new Americans may never have seen bananas in the old world, some American foods made token appearances there. In Josephine Cirella's small Italian town, they didn't know about bananas. "They just knew about the peanuts. And the only time you found them was at a feast, when they had a holiday feast, they had these vendors, and they would have these peanuts. And sometimes the peanut was that big you would find three nuts in the peanuts. Oh, everybody got so excited." To a child, nuts seemed to be a way to get to America. Remembering the Sicilian farm where her beloved grandfather was the foreman, Cirella said, "There was a lot of picking olives and picking almonds over there, and then they would have to crack them. And I remember going to a place where they would crack them with a little hammer, the almonds, all by hand, at that time. And then they got the kids to take them out of the shells and put them in a burlap bag. And we wanted to get into the burlap bag because they said that burlap bag went to America. Yeah. And they used to keep us kids, you know, busy that way."[216]

Oreste Teglia, forced to leave his pet dog "Pisarino" behind when as a twelve-year-old he sailed for the United States in 1916 from Ponte Bugiano in the Tuscany region of Italy, recalled, "I'd never seen a banana, but I knew of it because the people from America would tell me about it. They would come back to Italy once in a while, and they would talk about bananas. In fact when I got to this country, the first thing they gave me was a banana." Leaving Ellis Island, he remembered that his mother was given a lunch bag containing salami, bananas, bread and a few oranges, enough for the trip to Chicago. His banana experiences continued when he got a factory job stapling banana bushel baskets for the munificent sum of three dollars a week.[217]

As in Italy, the banana was part of American mythology in Scotland as well. Patrick Peak, one of nine children of the Peak family from Pomerdy, Scotland, gained an elevated status among the neighborhood kids when his mother decided that she and her off-spring should follow Patrick's father, older brother and one sister to the United States. "Gee all the kids were, 'Oh, he's going to America. He's going to America.' And back then, of course, when you came to America you could pick up gold practically on the backyard lots, you know. This is what the kids thought. And you could pick bananas and fruit off the tree. You know, life of plenty and ease. Which you find that isn't quite so. When you get over here you still have to work for a living." [218]

Peak's father, who had worked as a printer and transit worker in Glasgow, and his brother considered their job opportunities in New Bedford, Massachusetts to be worse than in Scotland. Their plans to return to Scotland were thwarted by Mrs. Peak's message to her husband that she had already booked passage for herself and the rest of the family. She hadn't, but she quickly did so. After a very stormy, six day early summer crossing and a six day stay on Ellis Island because the youngest brother had a rash on his face, the family arrived at New Bedford by ship with Patrick suffering from a terrible sunburn, an uncommon condition in Scotland.[219]

Some immigrants heard about bananas in Europe; some had their first banana encounter while still there. Sonya Thornblom Gillick came to the United States from Sweden in 1921. She was twelve. In Sweden she had eaten strawberries, pears, cherries and oranges. One day her father met a brother, a sea-captain, he had not seen in a long time. Her father came home with a green banana which he gave to Sonya.

> I was never without fruits when I was a youngster. I had never seen a lot of them. I had never seen a tomato and, of

> course, the only banana I ever saw was the green one that I got, my father brought. And he said, "It has to be yellow-you know, to get ripe to eat." So I put it in my bureau drawer. And, my mother said, "Every night when I lay in bed, I hear some gnawing, some noise. It must be a mouse someplace or a rat." And she saw little mounds of grain by the leg of this dresser where I had my banana. And it was a mouse had gone in there and bitten it into pieces. It was black when I got rid of it.[220] Getting a chance to eat a banana would have to wait awhile.

For Anna Kikta, who arrived at Ellis Island in 1928 as an eight-year-old, her first banana encounter was on a train traveling through France from her native Czechoslovakia. A man taking his family back to Canada said, "That's very popular in America. Everybody just loves bananas." Her father bought two from a peddler. "We peeled them and tasted them, and we all made faces, spit them out. And as the train started, we threw the banana out through the window. We didn't like them."[221]

Louis Stoller's mother gave him the nickname "Urtza," meaning a treasure in Hebrew. When his father departed for the United States leaving Louis, his mother and three sisters behind, he also became, at age eighteen, the breadwinner. In 1930 the family set out from Romania headed for Genoa, Italy to begin a nine day Atlantic crossing. Stoller and his family got only to Bucharest before potential disaster struck. They were detained there for five days when a medical examination revealed possible trachoma in Stoller's eye, a condition that would surely have prevented admission to the United States. Shipping companies were not inclined to transport immigrants who were likely to be sent back. The shipping company was responsible for the extra cost of transportation as well as a $100 fine for transporting any immigrant with a communicable disease. For Louis a treatment of alum and bluestone cleared the condition, but he still had to pass inspection by an American doctor in Genoa before boarding the *Conte Grande*.

It was in the ship's dining room that Stoller first met the banana. "And we never saw bananas in Romania. Where I come from in Romania, Soroki, there were no bananas. And we looked at it and looked at it until we saw somebody eating it. And we said, 'Well if they eat it, I'm going to eat it, fine.' I tasted it, and it was good. So I realized that was good."

One more surprise remained. When Stoller's ship arrived in New York City, a tag marked "EI" was placed on him, and he was loaded in a small boat. It must have been a relief when the boat arrived at "EI," Ellis Island.[222]

For immigrants traveling in steerage the food served on board ship was often different from their usual diet, and sometimes it was just plain bad. Herring, in a variety of forms, was usually on the menu. It was cheap. For two Polish emigrants, the mysteries of the banana were preferable to steerage class cuisine. Lawrence Meinwald, coming to America in 1920 at age six, decided to find food in the ship's first class garbage cans, but on one occasion "we all lined up and we were given bananas. I had never seen a banana in my life. So when I was given the banana, I bit into it skin and all until I was shown that you had to peel the banana."[223] Meinwald was neither the first nor the last to misjudge the banana by its cover.

As a sixteen-year-old young lady whose father died when she was eight leaving her mother with twelve children, Czeslawa Palenska Lutz set out for America in 1914. She traveled with some neighbors and cared for a neighbor woman who was sick the entire trip. Not liking the cooking on the ship, Lutz bought some oranges and bananas. "I buy a banana. I never see that in Poland. And I look like this, and I look. I said, "Gee, I want to know how you eat this.' And somebody said, 'Do like this.' And tell me open. I never liked banana, and I never like it now. I eat sometime because it's good for you."[224]

Madeline Polignano Zambrano not only saw her first banana on board the *Regina D'Italia*, but also shared the voyage with bananas in the ships' cargo. Although only four at the time, "I do remember they stopped in Portugal because my father took me off the ship, and we went on a rope ladder going down, while they loaded the ship with bananas. They came from Portugal. And we had our first bananas."[225]

Even if you came to the United States as late as 1952, you might not have seen bananas until you were crossing the Atlantic if you had been born in Germany during World War Two. Such was the case of brother and sister, Arthur and Hilda Broskas, who arrived at age nine and eight, respectively, at Ellis Island aboard the troop transport *General Stuart.* They came with their mother and father to join their paternal grandfather who had emigrated before the depression of the nineteen-thirties. Arthur, a Vietnam volunteer, recalled at the time he left Germany, "the American cowboy was a folk hero. He was the Romeo, the lover, and as a young man this is what I thought a lot of Americans were, cowboys, Tom Mix and the like. So, to me, that imagery stuck with me coming over, and I expected all Americans to be cowboys." Hilda remembered another American symbol. "Yes, I do remember seeing Mickey Mouse in Germany. And it was the black and white Mickey Mouse."

Having seen American magazines, Arthur remembers his desire to have a "big, shiny bicycle . . . . a red one was my big desire." But the family had to get here first. Arthur was instructed that in the presence of immigration officials he should be on his best behavior and never cough. If he did have to cough, he should have candy or an apple so he could pretend to be choking. Coughing, it was feared, could result in being sent back. On the ship the family was separated by gender. Since his father and brother were constantly seasick, Arthur, maintained communication with his mother and sister.

Recalls Arthur, "We didn't have any idea what Coca Cola or Jello or bananas were, until we came aboard ship." He bought a coke which he shared with his family. Hilda remembered her first exposure to Jello. Eating at a long table where the dishes were tied to the table, the first course served was a Jello salad. Because the ship was moving, the orange Jello was wiggling, a frightening experience to the uninitiated.

Ellis Island was also frightening. The family remained there for nine days until their father could convince immigration officials that he could support his family despite an artificial leg, the result of World War Two. He successfully made his case. When their train to Chicago struck a coal truck, they were unfazed; it was a minor incident compared to their war experience in Germany. The family eventually settled on a farm in northern Indiana where life was relatively peaceful and Arthur was fascinated by his grandfather's car.[226]

In 1914 Tessie Kalogarados Argianas was on her way to America with her mother and younger sister to rejoin the father who had left when she was only sixteen months old. No one stopped the nine-year-old steerage passenger as she made her way to the *Athena's* first class shop where candy was sold. Conversing in Greek, the clerk said, "Now, what you come? You got any money?"

Tessie relied, "When am I going to get any money? What do you think I'm going to America for? I'm going to have money when I go to America. Now give it [the candy] to me, and give me your address. I'll send you money." The ploy did not work.

When they arrived at Ellis Island, Tessie went to her mother to finance a different purchase.

"Mom, can you spend a nickel? I want to buy that long thing there."

"What is that," her mom asked?

"It's a fruit."

"All right, here get it."

Tessie went over and got the banana.

"Let me taste it first," said her mother.

Before she peeled the banana, she put it in her moth. Said Tessie, "Now, you know, how the heck you gonna eat the banana? This is funny. How you gonna eat the banana?"

"YUCK!"

"What's the matter? Isn't that no good, mom?"

"Ahh, you know, you know what it is?"

"Give it to me," instructed Tessie. "Look at the lady over there."

"So what?"

"You see how she peeled the banana? You got to eat the inside of the banana. You understand me, mother?"

"Here, you want to eat it, eat it. I don't want to eat that."

"Oh, is this good. Can I have another nickel?"

"No."[227]

Like Tessie, many immigrants found that arriving in New York was also arriving in the world of the banana. Such was the story of Sara Milanow Rovner when she arrived from Poland aboard the *Estonia* in 1925.

> When the ship docked [in Brooklyn] the sailors went down and they bought bananas. And I mean BANANAS, they bought plenty. We were all out on deck, naturally after so many, fourteen days on the water, we were on deck, and it was nice. The sun was shining. And they came over, and they gave each one a banana. We were standing, the whole crowd from downstairs, and everybody's looking at them. They don't know what it is. And I'm studying it. My brother says, "I'm not eating this." I says, "It's okay." I says, "If the Americans are eating it," I says, "I'm gonna eat it. I got to find out whether you eat the peel or you eat the inside. So I went and I broke it and I peeled it and I says, "That's what you eat! You eat the inside." And I bit into it. I didn't know what to do with it, to swallow it. And I didn't want to give them the satisfaction that I'm gonna throw it out So I'm eating it and I'm choking.[228]

If eating her first banana was a choking experience, her experience in Poland during World War One was pure hell. When the war began in 1914 there was no independent Poland. The area was controlled by Germany, Russia and Austria-Hungary. Rovner's father had left for the United States in 1912, and the family had virtually no contact with him from 1912 to 1920 largely due to the war. During the war when soldiers came to their house, they didn't know what language to speak, Polish or Russian. "They would shoot you right through the door," recalled Rovner. There was much fighting in the village of Dubno. The family slept in cellars, cemeteries, and barns. When they did return to their house, they found it occupied by Austrian soldiers. Along with two other families they were allowed to occupy one room.[229]

From the end of the war until 1925 Rovner's mother struggled to survive with her two girls and one boy. The father sent money which never arrived. Finally, in 1925 the family was able to emigrate after Rovner's father became a United States citizen. Not having seen her father for thirteen years, Rovner did not recognize him when they met at Ellis Island.

The kindness of the sailors who gave them bananas was matched by the Ellis Island employees who gave the children milk every afternoon. Because the family was "kosher," Rovner's mother objected to the children drinking milk less than six hours after eating meat. Rovner remembered the Ellis Island guard's response. "He says to her, 'Lady,' he says, 'This is not real milk,' because he wanted the children to have it because they were all undernourished. So he says, 'We take coconuts and we break them, and inside there is milk, and the milk goes in there and you can have it after meat.' So my mother said, 'What a country! Isn't that wonderful?' She believed him. She said, 'If that's the case, let them drink. It's good.'" After eight days at Ellis Island, the family set out for a new life in Boston.[230]

Many immigrants experienced not only the horror of the First World War but its confusion as well. "And one day we'd get up, it was called Poland; you'd get up the next morning it was Russia. But the day I was born it was Russia, so I'm a Russian," said Ethel Domerschick Pine who was born in the village of Neshevitz and came to the United States in 1921 at the age of eight. Her father, who had previously served in the Russian army, left home when Pine was six months old in order to avoid a second term of military service. Thus Pine remembers seeing him for the first time when their White Star line ship docked in New York in September 1921. Her father arrived in a small boat carrying a basket of fruit. Pine recounts, "I never saw a banana in my life. We had oranges, we had apples, we had grapes, we had pears. I don't even remember—tomatoes we had, but we didn't eat it. We gave it to the animals there. So what happened when I saw the banana? I liked the shape of it; I ate it with the skin. I didn't know. I took two bites. I says, 'Oh God, America what they pick on and eat.' And I took it and I threw it away. So my mother says, 'What's wrong with it?' I said, 'I don't like it. She said, 'It must be bad. If you're not eating it, it must be bad.' I was a good eater."

As was the case with many other new arrivals, Pine saw her first African-American when she arrived in New York. Her mother reassured her. " . . . . When I first saw, my mother says, ' They're just people like you. God loves them like the flowers. You have all kinds of flowers. You like them; they all smell nice. Even the stink weed is nice if you look at.' She says, 'Yes, and you have to respect people the same way. Don't look on their color. See what they are.'"[231] More than banana lessons were learned at Ellis Island.

George Monezis came from the town of Nenita on the Greek island of Khois. The one major crop on the island, and the only place in the world where it grew, was masticha, a flavoring made from the sap of a tree. While masticha is unusual, it was three new and unusual things that the ten year old George ate on his way to Ellis Island in 1922 that got him in trouble.

Traveling with his mother and sister, George ate his first Hershey bar in Naples and continued to consume chocolate bars during the trip. At a stop in Mallorca, he ate his first pineapple and banana. Then came the dreaded physical examination at Ellis Island. George tells the story.

> We walked in a great big room with cages all around it and everything else. And my face was all full of blisters, pimples rather. So the doctor looked at me. He says, "What's wrong with you?" Well, I didn't know how to speak English. He was an English[-speaking] doctor. So he told my mother, "We have to get an interpreter." So they went and got another young Greek doctor over there and he looked at and examined me and everything else. So he said, "What did he eat that he never ate before?" She says, "He ate banana, pineapple . . . . " He said, "Well, I'm going to keep you un-

> der observation. No more chocolate, no more bananas, no more pineapple." On the fifth day my face cleared up. He says, "That boy's allergic to chocolate or pineapple or bananas." He says, "We don't know, but we'll let you go." He says, "The boy's all right. So we'll let you go through."[232]

With great relief, George, with his mother and sister, set out to join his father who was working for the Weirton Steel Company in Weirton, West Virginia.

Rocco Morelli remembers "feeling a big shot that I was going to an adventure" on learning that he was leaving Aprigliano in Italy. His parents, both of whom had been in the United States from the time of their marriage in 1898 until 1907, had sold the house to pay for passage on the *Regina D'Italia*. There were thirteen in the party: Rocco, his parents, grandfather, grandmother and eight siblings. The twenty-eight day crossing in bad seas was not uneventful. Six days were lost in Gibraltar due to a coal strike. The constant hunger of a twelve-year-old caused Rocco to steal onions. Sadly, Rocco's older sister died of pneumonia during the voyage and was buried at sea.

When the party arrived at Ellis Island some of the children were ill and were kept in the hospital. Rocco recalled the hissing of the steam heat as all of his clothes were removed before his physical exam; they were returned later freshly laundered. Things began to improve.

> My uncles came to visit us about three days later that we were in Ellis Island. And one of them brought us a whole, now I know it was a hand of bananas, you know, and every banana was like that. They're nice and yellow. They picked me up like you'd pick up a football, you know, they were big guys; I was a little boy. So they told me, "Now you eat all the bananas because if we come back tomorrow and you didn't eat them, . . . you know." After they're gone I said to my grandfather, I says, "Nono, what do you do with these?" He says, "You eat them son, they're good. They're the American

> figs." He said to me, "You know how you like figs." I said, "Yeah." "Well," he says "that's how you'll like these bananas." I said, "All right." So I grabbed one and I went to put in my mouth. He said, "No," he said, "you got to take the skin." "Well," I said, "you said they were like figs. We don't take the skin off the figs." He said, "No," he says, "this you take the, you know—So," he said, "here, I'll show you." And he peeled it. He says, "Here, that's how you do it." So, I start to eat bananas. And that was the first time I ever saw, which, even today, if I have to eat ten, twelve bananas, I'll eat 'em, just like that. I mean, I don't care. I like them that much.[233]

Bananas served at Ellis Island did not come with instructions; but experienced hands were often there to offer advice. Salvatore Fichera, arriving in 1908, learned from his older brother. "They came around, they bring a sandwich, nice sandwich, a big—and they bring a banana, you know, bananas. I never see a banana in my life. I say, 'what's a this?' And I start eating without taking the skin off. My brother, because he was here, he says, 'What are you doing?' I said, "I don't know what's this I eat.' He say, 'Oh, no, no.' He says 'You've got to take the skin off.' That's the truth. So he show me; he take the skin off. 'Now,' he said, 'you can eat.' So we take a train to Bridgeport.

The sixteen-year-old got a job in Bridgeport, Connecticut working more than twelve hours a day and earning $3 a week as a shoemaker. He repaid a doctor in Italy who had loaned him the fare to come to America. Fichera returned to Italy in 1913 where he was taken into the army, but he eventually returned to the United States, now an experienced banana hand.[234]

Being greeted at Ellis Island by a father or other relative bearing bananas was a common experience. Being introduced to the banana almost became a rite of passage. Anna Tonnesen Bjorland arriving in 1925 was met by her father and cousin. "My cousin was standing on the side with a big banana, a big thing like this of bananas, something we never had in Norway.[235]

So too was the experience of Sarah Asher Crespi. "And the first thing, when we saw my father, of course, you know we kissed and that, and he gave us a banana, which we didn't have in Turkey, bananas. Maybe they had in Istanbul, but not in my town {Ankara]. And it was good. It was good[236]

If immigrants had not made an acquaintance with the banana on Ellis Island or before getting there, they were likely to discover the yellow fruit with the box lunch many purchased or were given when leaving the island. Sometimes the fruit wasn't yellow. One unlucky soul reported biting into a banana that was so green that it probably would not have ripened even if his train was going all the way to California. Elizabeth Milovsky Friedman's family left Ellis Island to take a train to Milwaukee where her father was to set up an upholstery shop. "They gave us boxes, lunch boxes, we didn't know what it was. I saw a man that took a banana out and he threw the banana away and he ate the peeling. He thought, you know there was some that never saw an orange or a banana."

Friedman's family came from economically comfortable circumstances in Kiev, Russia where her father had a decorating and upholstery business. But for a Jewish family, social conditions were poor and insecure. The family helped others go to America, but every time Friedman's family planned to leave, one of the children—there were ten—got sick or the passports expired. Friedman recalled experiencing a pogrom when the Jewish neighborhood was being looted and burned. Their Christian maid saved the family and house from attack by standing in front of the home with a cross. Finally, in 1914, before the outbreak of World War One hostilities, the family was able to make its way to America by way of Berlin and Bremen in Germany. They arrived in New York aboard the *George Washington*.[237]

Euterpe Boukis Dukakis' older brothers, already in the United States, paid the $80 fare for six people: Euterpe, her parents and three siblings. The ship, a former coal freighter, was very bad, and her father called Ellis Island a "place of tears or sighs." But when the family made its way to Massachusetts, they found life to be better than in their native Greece. Dukakis recalls, "The first real meal, other than the box which had been given to us on Ellis

Island, as we left—a sandwich and a banana. The likes of which we had never seen before and didn't know how to eat it. But my first real meal in the United States was lamb chops and French fried potatoes, which was, fortunately, kind of indicative of my life in the United States."[238]

It was only by a fateful decision that Jack Avruch got a chance to buy a banana in Boston. Avruch had a "good life" in the Ukraine. He took care of his father's rented cows while his father operated a creamery run. But all that changed when the "killing" of the Bolshevik revolution began. Avruch decided to enlist in the Red Army, the "only answer for Jews" he thought. On his way to enlist, Avruch met a known crook from his city. Armed with two guns, the man was looting and told Avruch, "take what you want." Avruch quickly saw that what is "good for the crook, [is] not good for me." He turned back, did not enlist, went to Warsaw, and made his way to the United States in 1920 at age nineteen.

After Ellis Island, Avruch journeyed by ship to Boston on his way to his sister's house. "When I went down from the ship, they were selling bananas. I bought two bananas. And I looked at them. I didn't know how to eat, to peel them or to bite them or to, what to do with them. So I saw what the other guy, how he did; he started peeling. So I ate the two bananas. That's all what I had [to eat since] a day before. So I was hungry."[239]

Settling into new homes and new jobs in America also brought new customs, new clothes and new foods. And yes, the banana. Such was the case with Verna Trill Horvath who arrived in 1913 from the village of Surnay in what was then Hungary, later Czechoslovakia. Her father had come two years earlier and had established a boarding house with ten boarders in Cleveland, Ohio. Along with her mother, brother and sister, and carrying there precious goose-down feather pillows, Horvath went to Cleveland by train. Her father planned a special surprise.

> So, and he said he's going to go to the market. We didn't live far from the West side Market in Cleveland. So he said he's going to the market. He's going to bring something that

> we've never eaten before. So he went and bought some bananas. And augh, we hated to disappoint him but it was the most awful stuff that I have ever eaten, you know, it was soft and smooth and mushy, and I didn't like it at all, but I swallowed it anyhow, just to please my father. But since then, I love bananas and every time I eat a banana, I think of that Saturday night when we arrived in Cleveland and I had a banana that I didn't like. And that's a good, long time ago and the memory always comes back.[240]

Horvath also remembered some tough times in Cleveland. Her father worked for the White Sewing Machine Company finishing cabinets for $1 a day. Later, he worked in a steel factory where he was crushed in an accident. While her father spent a long time in the hospital, there was little money for food. Horvath remembers eating meals of bread with mustard or bread with sugar. At age twelve she went to work in a business sorting strawberries—but not bananas.

Earlier arrivals sometimes tried to catch the rookies in the faux pas of incorrectly eating bananas. Elizabeth Repshis had such an experience after she arrived from Lithuania at age sixteen. Wearing shoes that weighed three pounds, she came with her parents in 1913. She went to Norwich, Connecticut to work in a paper factory. While there, her Polish landlady had a party. There was a big crowd including some Lithuanian boys.

> They had wine and beer and bananas and all kinds of food, you know. And she dressed me up, brassiere, you know, my breasts were just beginning to grow. I used to tie up so tightly so, I was ashamed, you know, but she threw my brassiere away and she said, "No in United States, the style that you have a breast." And my brassiere was lace, nice lace. Oh, I was so ashamed. They were drinking, of course, I wouldn't drink, you know, but what they wanna do trick on me, banana. I never saw banana, and I never saw the banana on a boat. So they gave me banana and everybody

> eating and they were watching me what I'm going to do with the banana. But I didn't eat. I was waiting if somebody else started! I saw them peeling; and I peeled it. They didn't catch me.

Repshis also tasted her first ice cream and sat in her first rocking chair at the party.[241]

Esther Shapiro Gologor, a Jewish immigrant from Palestine in 1921, also experienced ice cream and bananas for the first time in America. Her uncle, who had arrived in America a few months earlier, brought both items to their apartment. It was a real treat for the twelve-year-old.[242]

After being left behind with an aunt in Austria when her mother came to the United States, Marie Tancibok Vitosky was a happy nine-year-old when she was reunited with her father at Ellis Island in 1905. He brought her bananas. "I tasted them. I thought, Oooh, I didn't like bananas. Ach!" Then came a discovery she did like. "And then he showed me the machine where you put the penny in for the chewing gum. We never had chewing gum. We didn't know what chewing gum was."[243] Ah America, the promised land of bananas. ice cream and chewing gum.

For some immigrants, bananas were a mysterious treat no matter where they were found. Mary Masare Thome came to the United States in 1909 from the part of Austria-Hungary which became Czechoslovakia after World War One. She, her mother and younger brother followed her father's emigration by two years. The family settled into a two room row house in Kenosha, Wisconsin where both Thome's father and mother worked in a factory. "Do you know, one thing I remember in Kenosha. Somebody had dropped a banana peeling. And I didn't know a banana from Adam I guess. And I remember scraping the inside of the banana on my teeth to find out what it tasted like. And, I think after that I did it quite often, because I don't think I got bananas at home. Not that we ate very little, but we ate very economically, probably mostly bread and beans." The relatively austere quality of immigrant life is illustrated by the family's housing when they moved into a rented house in Racine. There were five rooms: four bedrooms and a kitchen. They took in sixteen boarders who slept two in a bed with four beds in a room. Thome shared a third room with her mother and brother.[244] Such accommodations, while not uncommon, gave new meaning to "huddled masses."

For some immigrants, bananas became more than a new food; they became a job, if not a career. The stories of Samuel Zemurray, who rose to direct the fortunes of the United Fruit empire, and the Vaccaro Brothers and their New Orleans based Standard Fruit Company, have been recounted earlier.

Andrew Preston, one of the founders of United Fruit, noted that throngs of immigrant banana peddlers and some dealers, mostly Italian, Greek and Slavic, were selling amazing tonnages of bananas to a rapidly growing population of immigrants. They operated much in the tradition of the "Yankee peddler." Preston quoted, or perhaps even wrote, a lecture entitled "Lessons of the Banana Man."

> Almost anybody can do things with plenty of capital, but it takes a real merchandiser to do business on a scanty amount of cash. The banana peddler loads his cart in the morning with, say, ten dollars worth of fruit. He returns at night with an empty cart and fifteen dollars in cash. He turns his capital maybe three hundred times a year. On a gross yearly business of $4500, he makes some three hundred separate profits of thirty-three and a third percent, and all on an original investment of ten dollars plus the cart and his time.[245]

Marianthe Dimitri Chletsos saw peddlers first hand from her aunt's house on East Twenty-eighth street in New York. "It used to be, uh, the wagons with the horses selling fruit. They'd sell bananas and all kinds of vegetables. And ice, they used to holler everything, 'Oh, bananas, oh, apples, or 'ice!' Screaming. If you need anything you used to come out the window and wave to them. They would send it up."[246]

In the early twentieth century many banana peddlers were immigrants. (Courtesy of Library of Congress, Prints and Photographs Division)

When Rose Ganbaum Halpern arrived from Russia in 1923 with her father and four siblings, the family was met by her aunt and uncle. The uncle was in the banana business in the Bronx. Her father was given a stand to sell bananas.[247]

Although bananas were virtually unknown in her native Albania, Pauline Stevens Curtis' father entered the wholesale and retail banana business in Rockland, Maine in the 1930s. The bananas came by boat from Boston. Curtis remembers her father carrying stems of bananas on his shoulder and loading them in wagons. In what was a forerunner of today's ripening rooms, the bananas were hung in the basement where the family had to tend gas stoves so the bananas would not ripen too fast or too slow. Horse wagons, later replaced by trucks, were used to make banana deliveries.[248]

An earthquake in Sicily killed the owners of the big estate where his father was the manager, so John Santo Fieramosca's father emigrated to the United States.

John, his mother and sister followed nine years later. Fieramosca was a sick thirteen-year-old for all seventeen days of the trip in 1924. When his mother, also sick the entire voyage, was released from the Ellis Island hospital, his father took them to a house in Elizabeth, New Jersey.

> Well then my dad got a permit for me to work. I was fifteen years old. And I kept working ever since. Well, we did everything we could find. I sold newspapers, I worked in a nut factory and I made banana bushels at the market, anything you could find to make an honest buck. You know those bushels that you put the bananas, the bunch of bananas in. I would just staple the bushels together. We kept giving our fathers our earnings until we were of marriageable age. There was a market right near where we lived. You know, they sell fruits, everything. And there was an Italian man, he used to sell bananas. And I knew this man. This man's brother worked for my father in Italy, you know? So I used to hang around there, a couple of hours everyday selling bananas. And I learned to say, "BANANAS, two dozen for a quarter." Now you get one banana for a quarter. Then I started to work with another guy. You know, part time jobs, small jobs. He had a horse and wagon and he used to sell vegetables all over, house to house. And he used to give me a little basket. I hang it on my arm and put like six bananas, a little bit of everything, and you go to a house and you sell them and they pay you—fifty cents, depends how much they spend. That was good. Then I started to work in a factory.[249]

The factory job was followed by selling cars and then limousine work until the depression. That was followed by starting a garment manufacturing business in Elizabeth, New Jersey where he made uniforms during World War Two. Eventually, Fieramosca became a successful thoroughbred horse breeder in New Jersey. Not bad for a young banana peddler.

Perhaps the immigrant and the banana had much in common. Neither was native to the United States; they were both immigrants. While no one is quite sure of the origin of the term "greenhorn," it was often applied to the new arrivals. Many an immigrant child suffered the "greenhorn" taunts of school mates. Perhaps one immigrant summarized the situation best. "When we got to my aunt's house, there was a houseful of people who had come to see the Greene —G-R-E-E-N-E. When you first came to this country you were called a greenhorn. So in Yiddish that make it the Greene, those that are green, 'til you ripen and mature." Just like a banana.

Some ripened faster that others. Some assimilated faster than others. Sometimes the banana was part of the process as was the case with Betty Dornbaum Schubert an immigrant from Rumania in 1911. She was only seven. Her father had died before she was one, and her mother had to struggle to bring Betty, her sister and grandmother to the United States. Living in a crowded tenement in New York City, Betty had to sleep with her two feather pillows on two chairs pushed together in front of the coal stove in the kitchen. With her mother working nine to eleven hours a day, Betty was often left to her own devices.

> Oh, being I was left by myself, I wanted to be an American—very fast. So I went with the bad bunches, cliques, in the street. So, I joined the cliques. If they stole a banana or if they just did something, if they shimmied up the pole, I did the same thing. And we had the horses draw the trolley cars, so if they go up the lamp posts and throw a bag of water at them, or flour, I had to do the same thing to be an American.[250]

While stealing bananas may have been part of assimilation for the young, in the eyes of many immigrants, liking bananas was a key part of becoming an American. Some learned to like bananas faster than others. The story of Roslyn Bresnick-Perry who arrived from Poland in 1929 is illustrative. Roslyn was sick during the entire crossing. She threw up every time her mother tried to give her herring during the ten day trip. When they landed, Roslyn did not recognize her father who had arrived six years earlier. He was "fat, red-faced," not the handsome man she had always seen in the pictures.

> And we go into the cab, and my father's holding my mother around, and with the other hand reaches into his pocket and he pulls out a banana. And he says to me, "You know what this is?" I say, "No." And he said, "This is a panana." He couldn't say banana. He said panana. He says, "This is a panana." He said, "Do you know, did you ever see it before?" I said, "No." He said, "In America, all the children eat pananas." And he says, "You want one? You want some?" Now, I didn't feel like eating anything, but I felt I already had done terrible things to my father. I wanted him to like me. After all, he was my father. So I said, "All right." He said, "But now I'll show you." He says, "I'll show you how to eat it." He says,"You see, you hold it up like this, then you take it like this and you peel it down with the shell hanging down, like an umbrella," he says. He says, now he says, "Take a bite." So I take a bite and I started to chew the banana, and it had this, I'd never eaten a banana in my like. I never saw a banana, it had this sweet, cloying taste, you know. The consistency was something I was not used to. It somehow reminded me of my nausea. And I start to gag, and my mother put a handkerchief under my mouth, and I threw up again, because my stomach was still all upset, and the banana really got me right here. And my father looked so disappointed. And my mother cleans me up, and he said, then he cheers up. My father, a cheerful man, he says, and here he says, and he points with his finger, like up, up to heaven, you know, with his finger. He says, "You know, some day," he says to me, "some day, you're going to eat a whole banana, and then you'll be a real American girl." Well, I want you to know it took me a very long time to be able to

> eat a banana, and bananas are still not my favorite fruit. But now I eat them because at my age the doctor says I need the potassium.[251]

The banana seems to enjoy a special place in the collective memory of those who have recorded their experiences for posterity. But , of course, it is more than banana stories that Dr. Levine and her colleagues have recorded at Ellis Island and around the country. As Josephine Cirella said at the conclusion of her recollections, "You brought me back, all the way back, and a lot of memories, and they were good memories. As I've been talking to you, I've been picturing everything that I told you in front of my eyes. And even my grandfather, when he left us on the train, how much I cried. You brought me back there, so I'm happy for that. At least I remember."[252] As Dr. Levine has stated, "I consider my job making people feel honored. The tapes are important for the family and for the library." And the tapes are there for all to hear both now and in the future. If you visit Ellis Island, stop at the Oral History Library where you can use one of the twenty computers where you can select a tape to hear while also viewing the printed transcript. You can make your choice by name of the immigrant, country of origin, year of arrival, ship or by one of several major topics. You might just hear a Swedish immigrant say, "I didn't know what it was. Ooh, I thought it was awful. But I learned to eat them when I came to America. As long as there was ice cream and bananas, I was staying."[253]

CHAPTER 5

Banana Republic Politics

"This entire base is like the big white plantation house and the rest of Honduras is one big Chiquita banana field. As a Latino, frankly, I'm embarrassed to say this is still a banana republic."

U. S. Army Sgt. Noel Sánchez, speaking in the 1980s of a U. S. military base in Honduras.[254]

The morning of June 18, 1954 was not a good time for Jacobo Arbenz Guzman, the President of Guatemala. The forty-one year old president, in his fourth year of a six year term, was only the second democratically elected president in the Central American country's one hundred thirty-three years of independence. A C-47 transport plane had just swept over Guatemala City dropping thousands of leaflets demanding Arbenz's immediate resignation and threatening to return in the afternoon to bomb the National Palace and the arsenal if the president did not step down. Twice in the weeks before a plane had dropped leaflets warning the Guatemalan Army that Arbenz planned to replace the army with a citizens' force if the army did not rise up against the president. Arriving in his office, Arbenz was informed by aides that a plane or planes had strafed buildings and gas storage tanks in the Pacific coast city of San Jose. He also received reports that a long threatened National Liberation offensive had begun. Insurgents commanded by rebel leader Colonel Castillo Armas had crossed the frontier from Honduras and had overrun a frontier post at La

Florida. While the reports were sketchy, they were ominous. Newspapers in neighboring Honduras were reporting that chartered DC-3s were airlifting rebel troops from Tegucigalpa to the Guatemalan border. Arbenz was grim; he had cause to be. During the past nine years of democratic rule under Arbenz and his predecessor, Juan José Arévalo, the government had survived over thirty attempted coups by right-wing Guatemalans.[255]

Through most of the twentieth century revolutions and coups in the "banana republics" of Central America were as common as corn in Iowa. What was not so common about this 1954 episode in Guatemala was that it was a case of one democracy overthrowing the democratically elected government of a neighbor. Although not generally known at the time, but perhaps suspected in some quarters, the "National Liberation" offensive was actually a covert operation of the United States Central Intelligence Agency instigated by the United Fruit Company to protect its interests in Guatemala. In other words, it was "banana politics." Although the United States had politically meddled in the affairs of "banana republics" before, this case was different.

The United States government had pulled the "bananas out of the fire" on previous occasions. During the 1930 Louisiana campaign for the United States Senate, Huey Long, the "Kingfish," charged his opponent, Joseph E. Ransdell with having made a corrupt arrangement with a New Orleans fruit importer—in Long's words, "a banana peddler." The importer of course was Samuel Zemurray who had given Ransdell's nephew a job allegedly to have the Senator induce the War Department to dispatch troops to Central America to protect Zemurray's holdings. Long's only foreign policy issue seemed to be that the United States should not send troops to Latin America to protect the interests of Samuel Zemurray and the United Fruit Company.[256] In 1954 it was largely the United Fruit Company that induced the government to direct the CIA to bring down the Arbenz government.

The reason for United Fruit's discontent can be traced to 1944 when the "October Revolution" led by schoolteachers, students and a rising middle class deposed the fourteen year military dictatorship of General Jorge Ubico. The general, who fancied himself another Napoleon and was accompanied by an official biographer everywhere he went, had suppressed all dissent and disbanded all labor unions. In 1945 in its first legitimate election, Guatemala selected a previously exiled teacher, Dr. Juan José Arévalo Bermejo, to lead the country. The Arévalo government eliminated the secret police, guaranteed freedom of speech and the press, repealed unfair labor laws and gave workers the right to organize and strike. The government's role began to shift from a pro-capital to a more pro-labor position. As the largest employer and landowner in the country, the United Fruit Company began to get nervous.[257]

Arévalo's successor was elected in 1950 with sixty-five percent of the vote. He was Jacobo Arbenz Guzmán who was also a leader of the 1944 revolution. Land reform, the biggest promise of the revolution, still needed to be fully addressed. A 1950 census showed that 2.2% of the landowners still owned seventy percent of the nation's land available for farming. Less than one fourth of the land in the hands of the large owners, about four million acres, was under cultivation. United Fruit alone owned over a half million acres; eighty-five percent of it was uncultivated. A priority of the Arbenz government was to give land to Guatemala's landless peasants. It was obvious from where much of the land would come.[258]

An agrarian reform act known as Decree 900 was adopted on June 27, 1952. In two separate decrees 209,842 acres of uncultivated United Fruit land was expropriated in March 1953. In compensation the government offered bonds in the amount of $627,572, exactly what the company in its tax declaration said the land was worth. Everyone knew that the company had historically undervalued its land to reduce its tax liability. In April of 1954 the U. S. State Department, not United Fruit, lodged a formal complaint with Guatemalan authorities demanding $15,854,849 in compensation for the company. Thus the State Department was demanding about $75 per acre for land which the company had purchased for $1.48 per acre twenty years before and for which Guatemala was now offering about $2.99 per acre. The Guatemalan foreign minister refused to accept the note and branded it an attempt to meddle in Guatemala's internal affairs.[259] Little did he know how prophetic his position was; United Fruit and powerful forces were already at work to end the "Guatemalan problem."

Two dynamic personalities with innovative talents, both foreign-born and self-made, had been concerned about United Fruit's position in Central America for some time. Samuel Zemurray, the head of United Fruit, and Edward Bernays, widely regarded as the "father of American public relations," saw trouble brewing in 1947 when the Guatemalan Congress enacted a labor code which permitted workers to join unions. An aggressive Bernays launched a counterattack against the Guatemalan government. Bernays used his close contacts with the media to promote the idea that Guatemala was leaning toward communism, an idea that attracted official and public attention in the age of the Korean War and domestic McCarthyism. To Bernays the land takeovers that began in 1953 confirmed that the Arbenz regime was communistic. Senator Henry Cabot Lodge, a large United Fruit stockholder and the senator from the company's home base of Massachusetts, denounced the labor code on the floor of the U.S. Senate. At least five press junkets, designed with the "precision of a space shot," took editors and reporters to Guatemala where they saw and heard what the company wanted them to see and hear. As a United Fruit public relations man observed in this case of press manipulation, the "victims proved so eager for the experience."[260] Thereafter, public interest in communist influence in Central America skyrocketed.

United Fruit also put a Washington "insider" to work on its Guatemalan problem. He was the well-connected lawyer Thomas G. Corcoran, "Tommy the Cork," who had been hired as Zemurray's "personal counsel" and lobbyist in 1947. As one of Franklin Roosevelt's original "brain trusters," Corcoran had close ties with Washington Democrats and close friends in the CIA. Corcoran worked with William Bedell Smith, a former CIA man, in helping plan the coup. Smith suggested that he would be interested in the presidency of the United Fruit Company, and he was later named to the Board of Directors. Smith was the Undersecretary to Secretary of State John Foster Dulles who was known as an ultimate "Cold Warrior." John Foster Dulles and his brother Allen, Eisenhower's Director of the CIA, were both stockholders in United Fruit. Anne Whitman, the wife of United Fruit's Public Relations director Edmund Whitman, was Eisenhower's personal secretary. With connections like those Corcoran's job could not have been too difficult. E. Howard Hunt, later of "Watergate" fame, was the propaganda and political action officer of the covert action against the Arbenz regime. Hunt credits Corcoran's lobbying for the National Security Council's decision to proceed against Arbenz.[261]

The view that the Guatemalan government under President Arbenz was leaning toward communism was probably overblown because of press manipulation and the tendency in those days for Americans to fear that "reds" might be lurking under their beds. In actuality, communists held only four of the fifty-one seats in the 1953-54 Guatemalan Congress. No communist had ever held a cabinet position although seven or eight held sub-cabinet posts. Their greatest influence was within the labor unions. Many groups in Guatemala were far more powerful than the communists. Arbenz was primarily a nationalist and accepted communist support when necessary.[262]

The openness of Guatemalan society from 1944 to 1954 attracted exiles not welcome in the other "banana republics" of Central America. Among those drawn to Guatemala was a relatively unknown young Argentine doctor, Ernesto "Ché" Guevara. In a letter he wrote to his aunt in Argentina in 1953 Guevara remarked that he was much concerned about the "dominions of the United Fruit." He was concerned of how "terrible the capitalist octopuses are," and he swore before a picture of Stalin that, "I won't rest until I see these capitalist octopuses annihilated." Guevara later became Fidel Castro's "right-hand man" in the Cuban Revolution. Castro was a native of Cuba's Mayarí province which was a virtual vassal state of the United Fruit Company. Castro's father leased land from "the Company" and was compelled to sell his sugar cane to United Fruit mills.[263] The presence of Ché and other leftists in Guatemala probably accelerated the impression of Marxist influence in Guatemala.

Early in 1954 the CIA was authorized to proceed with a plan to depose the Arbenz government. "Operation Success" was essentially a plan of psychological warfare not unlike the CIA led overthrow of Iranian Premier Mohammad Mossadegh in 1953. The operation was designed to psychologically prepare the minds of the target government and population. This was to be accomplished by clandestine radio broadcasts and airdrops of propaganda leaflets. The Guatemalan people and the army had to be convinced that Arbenz could no longer control the country. While the operation was not primarily military, a military force had to be raised to invade from a neighboring country. The idea was to have a sudden show of an apparently massive force.[264] The plan also anticipated the neutralization "through Executive Action"—assassination—of fifty-eight Guatemalan officials. The spy agency planned to install an "authoritarian" government in place of the functioning democracy for the sole purpose of protecting U.S. interests which included those of the United Fruit Company.[265] The coup managed to avoid the assassinations, but after a new repressive regime was installed large numbers of labor leaders and student activists later lost their lives.[266]

"Operation Success," with its own funds, communication center and chain of command, set up shop in Opa Locke, Florida in a barracks of a half-closed Marine air base. Head of the operation was Albert Haney, later described by another CIA man as "our Oliver North kind of out-of-control type."[267] There under the direction of E. Howard Hunt a cadre of Guatemalan newspapermen were turned into propagandists as they prepared newspaper articles, pamphlets and leaflets in Spanish. A series of broadcasts was also prepared for the operation's D-Day. When they threatened to strike due to their unaccustomed celibacy, Hunt arranged to have girlfriends flown in without immigration formalities and installed in a safe house . There was no further strike talk.[268]

The "Army of Liberation" was to be headed by Colonel Carlos

Castillo Armas, a selection made with considerable input from United Fruit. Armas had been exiled in 1950 after an abortive coup attempt. Now he and his army of about one hundred fifty men were positioned at a training base in Nicaragua provided by that country's dictator Anastasio Somoza. The army included some Guatemalans opposed to Arbenz, some other Central Americans and American mercenaries. Americans were recruited for the rebel air force which was composed of an assortment of thirty planes—an overwhelming force by Guatemalan standards. The planes were acquired through phony deals that included, among other devices, a "Medical Institute" which received the planes as a tax-deductible charitable donation and then sold them to front corporations engaged in "aerial photography, crop dusting and recreational aviation."[269] If the I.R.S. had investigated the operation, they might still be conducting an audit to trace the money flow.

The key radio component of the plan was known as "La Voz de Liberacion" (The Voice of the Liberation). The programs were to appeal to patriotism and the base values of the society. "Trabajo, Pan y Libertad" ("Work, Bread, and Liberty") was the slogan adopted.[270] Apparently, the leader of the liberation army, Colonel Armas was on a different wavelength. He adopted the motto "God, Fatherland and Liberty."[271] At least they agreed on liberty. The main radio base was set up in Nicaragua, and additional transmitters were located in Honduras, the Dominican Republic and even in the U. S. embassy in Guatemala City.

The job of psychologically readying the Guatemalan population, and for that matter the government and army, for the liberation began on May 1, 1954 when La Voz de Liberacion took to the air. Since no one would listen to an underground radio station unless they knew it was on the air, the station's debut was advertised in the Guatemalan press. The broadcast which had advertised that it would have popular Mexican entertainers instead denounced the president and said he would be overthrown by rebels. Arbenz made a speech on Radio Nacional the next day denouncing the rebels. CIA jammers drowned out Arbenz's speech. The main task of the rebel station was to convince the country that

Arbenz was not in control. Appeals to the Guatemalan air force prompted one defection. After that, Arbenz grounded his air force, all six pre-1936 training planes of it, for the duration of the crisis. Doubting the loyalty of his army, Arbenz made plans to arm the peasants and trade unions. The army began to question the intentions of the president when the rebels dropped leaflets calling attention to the president's plans to arm civilians. This then was the situation on June 18, 1954 when Colonel Castillo Armas, riding in an old station wagon, crossed the Honduran border. His one hundred fifty soldiers followed in cattle trucks.[272] CIA officials apparently did not want Castillo Armas to be too well equipped; at one policy planning session these officials debated whether the rebel leader's Piper Cub aircraft should be equipped with new or retread tires.[273]

The next day La Voz de Liberacion broadcast reports of phony troop movements including thousands of rebels crossing the border. The CIA launched occasional bombing and strafing raids on military bases. Sometimes the pilots dropped smoke bombs and even empty soda bottles. The Guatemalans called these planes "sulfatos" (laxatives) because of the effect they had on government officials. Using Guatemalan army radio frequencies, the CIA began broadcasting false commands which further confused the situation. An effort to put Guatemala's case before the United Nations was thwarted by Henry Cabot Lodge, the U. S. Ambassador to the UN.[274] On Sunday, June 27 the rebel radio was announcing that large columns of rebels were approaching the capital city. In fact Armas and his one hundred fifty men were lounging in Esquipulas near the Honduran border. The CIA did all that it could to prevent press access to Armas. One resourceful female British reporter rented a mule, rode into the rebel camp, and got an interview. When the *New York Times* got word of the British exclusive, it wired its stringer in Guatemala City, "Get off ass—get on burro."[275] As it turned out, the United Fruit Company was one of the best albeit not always accurate sources of information.

On the night of June 27 President Arbenz went on the national radio and announced that he would resign in order to bring peace to the country. The CIA operatives were stunned by the ease with which their plan had succeeded. The whole exercise seemed to demonstrate a new version of an old adage, "The mouth is mightier than the sword." After several days of political maneuvering with the Guatemalan army orchestrated by American Ambassador Peurifoy, Colonel Armas and his troops flew into Guatemala City. Armas was installed as the president of Guatemala amidst a celebration accompanied by firecrackers supplied by the CIA. Arbenz was allowed to go into exile and eventually made his way to Mexico.[276] But all was not a total success. Against orders, a CIA pilot dropped one bomb that sank a British ship in the Pacific port city of San José. The ship carried the dangerous war materials of coffee and cotton.[277]

United Fruit got its land back and a reduced tax, but in the long run the 1954 coup in Guatemala did not produce winners, only victims. Five days after Arbenz resigned the U. S. Department of Justice sued United Fruit for antitrust violations. For some time the Department had been examining the company's monopoly of banana exports from countries like Guatemala. The company attempted to discredit the Justice Department by suggesting possible communist influence in the Department. Bernays promoted a series of editorials questioning why the government that had received help in fighting the "Red Menace" in Guatemala should suddenly turn against its patriotic friend, the United Fruit Company. The company produced a short film, *Why the Kremlin Hates Bananas*, which tried to make the same point. The film had a short life; a more sensible company leadership later withdrew the film and destroyed all of the prints.[278] In 1958 the company agreed to a consent decree and reduced its holding in Guatemala. In 1972 the company sold its remaining landholdings in Guatemala to the Del Monte Corporation and withdrew from the country.

In 1961 the United Fruit company was called upon to return a favor when the John F. Kennedy administration authorized E. Howard Hunt and the CIA to launch the abortive Bay of Pigs invasion to topple Cuban leader Fidel Castro. Undoubtedly, the successful operation in Guatemala served as a model. Dealing directly with Bobby Kennedy, United Fruit was called upon to provide two freighters for the operation. After the invasion failed, the logs of the two ships were sent to Washington and later returned to the company sealed in wax.[279]

For Guatemalans the 1954 coup set off civil wars that did not end until September 1996. More than one hundred thousand people were killed and the government changed frequently as the army, leftist guerrillas, death squads and the wealthy local oligarchy struggled to control the country. By the time the government and the rebels signed a peace accord in 1996, most of Guatemala's 10.5 million people could not remember what started the conflict. But in early 1997 the United States sought to enlist Guatemala in another war—the war on drugs. The U. S. Drug Enforcement Agency planned to train and equip the Guatemalans to locate and interdict drug trafficking. Some saw this as repeating a mistake. As Harvard scholar Jennifer Schirmer noted, "These guys just don't get it."[280]

By 1968 the United Fruit Company was seventy years old and had survived every type of upheaval. It had survived thousands of acts of God in the form of droughts, hurricanes, blights and floods. It had survived hundreds of political revolutions, two world wars, depressions, strikes, protests and expropriation. It had outlived many of its competitors, but it was on the verge of a new era. A year later control of the company was bought by Eli Black, an ordained Orthodox Jewish rabbi and descendent of ten generations of rabbis and scholars. Starting with a small bottle cap company, Black combined a group of small manufacturing companies into AMK Corporation and acquired John Morrell and Co., a meat packer. In 1970 AMK and United Fruit became United Brands with Black as chairman and president. United Brands became an empire of bananas, cattle, root-beer stands and ice cream parlors when A & W International and Baskin-Robbins were acquired. Black immediately set out to change the image of "el Pulpo." He redoubled previous United Fruit efforts to improve health and education conditions on the company's Central American plantations and even began to turn over many of its land holdings in Costa

Rica and Honduras to the governments. He opened new plantations in the Philippines and began negotiations to open markets with several communist governments in Eastern Europe. In 1970 Black personally negotiated on behalf of a United Brands lettuce subsidiary with Cesar Chavez's United Farm Workers to reach the union's first contract with a lettuce grower. It was a new era for the banana company.[281]

On February 3, 1975 Eli Black decided to leave the company. At 8 A. M. on that day he used his brief case, loaded with books, to smash the quarter-inch plate glass window of his office in what was then the Pan-Am building in New York City. He then plunged forty-four floors to his death, falling on the roadway before horrified motorists. Investigators found that before jumping Black had removed some of the broken glass from the three by four foot hole in the office window. As Detective Duffy observed, "he apparently didn't want to cut himself." The death was ruled a suicide. Black left no note.[282]

The suicide was initially attributed to business pressures. The company had lost money in 1972; Hurricane Fifi caused $20 million in damage in Honduras. Banana exporting countries were demanding new higher export taxes, and there were millions in cattle feeding losses.[283] Not known at the time was what became known as "Bananagate."

The origins of "Bananagate" can be found in the 1974 attempt by seven Latin American countries to form UPEB, Union of Banana Exporting Countries, which was inspired by OPEC, the oil cartel. UPEB proposed an export tax of one dollar for each forty pound box of bananas exported. The companies protested and threatened to withdraw their operations. The plan died partly due to a glut of bananas. Costa Rica, however, did set a tax in 1975 but reduced it to twenty-five cents per box. In 1974 Honduras passed a fifty cent per box tax, enough to cost the companies dearly. When the Securities Exchange Commission investigated the Black suicide, it discovered that United Brands had paid a bribe of $1.25 million to Honduran President Oswaldo López Arellano and was to follow with another $1.25 million the next year. After the bribe the Honduran tax was reduced from fifty to twenty-five cents per box. In addition it was discovered that United Brands had paid about $750,000 in bribes to an Italian official since 1970. None of the bribes could have been paid without the knowledge or approval of Black. While it was not illegal at the time for American companies to make such foreign payments, it was illegal for companies to juggle their books to hide such bribes from their stockholders.[284] Greasing government palms was still part of doing business in the "banana republics,"and Black's efforts to improve "el Pulpo's" image had suffered a blow.

Along with "Bananagate" and the Black suicide in the early 1970s United Brands also lost its leadership in the U. S. banana market to the Dole brand owned at the time by Castle & Cooke Inc. United Fruit once had a dominant eighty percent control of the U. S. market, but by 1977 it had only a thirty-three percent market share although bananas accounted for seventy percent of United Brands' profits.[285] After several changes in leadership, United Brands was due for another shakeup.

In 1987, the company came under the control of Carl H. Lindner, a Cincinnati businessman whose career began when he dropped out of school to run his family's ice-cream store. Lindner built the business into a chain of stores known as United Dairy Farmers. From the dairy business Lindner moved into finance by acquiring savings and loan associations in Ohio which then became the springboard to the insurance business and a corporate conglomerate known today as American Financial Corporation with assets worth over $14 billion. Along the way Lindner's American Financial employed Charles H. Keating, Jr. who purchased a home building division of Lindner's empire and gained notoriety in the 1980s as the man behind the multibillion-dollar failure of Lincoln Savings and Loan. Lindner developed a Wall Street reputation for buying undervalued and financially troubled companies at bargain prices. He was regarded as an important customer of "junk bond" king Michael Milken.[286] At the 1976 annual stockholders meeting of United Brands a shareholder asked why Mr. Lindner had invested in United Brands to the point where he owned

20.5% of the outstanding shares. Chairman Max M. Fisher's bland response was, "Mr. Lindner has asked me to say he is intrigued with the possibilities of this particular company."[287] In 1987 Carl Lindner took over United Brands and in 1990 changed the name to Chiquita Brands International.

The company that Lindner took over in 1987 had many intriguing problems along with the possibilities. One of the problems was the Honduran "banana war" of 1990. In 1989 an Irish fruit company, Fyffes Plc, hired Ernst Otto Stalinski to secure banana suppliers from Honduras and break the Chiquita monopoly in that country. Until 1986 Fyffes was a Chiquita subsidiary. Stalinski approached the Echeverri family, one of Honduras' biggest independent growers, and offered $2.5 million more that what the company was getting from Chiquita. Fyffes was offering about $4.40 per box while Chiquita was paying about $3 a box for the bananas. The ensuing hostilities included machete-wielding peasants confronting gun-toting uniformed men blocking the shipment of Fyffes bananas. In one incident armed men boarded a ship at gun-point and destroyed one hundred thousand boxes of Fyffes bananas. A train loaded with Fyffes bananas was derailed. The conflict culminated on a Saturday afternoon in April of 1990 when uniformed men and a Chiquita attorney tried to arrest Stalinski in his hotel room in San Pedro Sula, Honduras. Stalinski escaped through a rear basement exit. He later accused his pursuers of attempting to kidnap and murder him. Several weeks later in a deal brokered by the Honduran government and the European Community, Chiquita and Fyffes signed a cease fire agreeing to peacefully coexist in Honduras.[288]

Chiquita's problem did not quite go away. In 1995 Otto Stalinski, the former Fyffes operative, filed suit in Honduran civil and criminal courts against eight individuals allegedly connected to Chiquita. The charges included attempted kidnapping, blackmail, coercion, housebreaking, extortion and attempted murder. Chiquita officials denied any connection to the individuals charged.[289] The case was later dismissed, but someone's bananas had been bruised.

Along with the commercial war, strikes, land disputes and other problems reduced Chiquita's role in Honduras. In 1987 the company exported thirty-two million forty-pound boxes, but the number dropped to thirteen million boxes by 1995. There were banana worker strikes in Honduras every year between 1990 and 1994. Chiquita sold off some of its land and began to contract for more bananas from independent producers.[290] The closing of some Chiquita plantations led to land disputes and the uprooting of hundreds of families. At Tacamiche plantation in February, 1996 army troops, police and Chiquita work crews evicted one hundred twenty-three families. Crops were destroyed and the wooden cabins and three churches in the settlement were razed. The families contended that their families had lived on the land in the 1920s before Chiquita's original purchase of the land for $1 in the 1930s. One human-rights investigator claimed that records at the Honduran title registry showed that Chiquita has sold the land three times since its original purchase. Chiquita claimed that the land was registered in the name of its subsidiary, the Tela Railroad Company. The President of Honduras supported Chiquita's claim.[291]

Land disputes are not the only problem the banana companies face with their workers. In 1997 twenty-five thousand workers from twelve different nations were suing three banana companies and chemical giants Dow Chemical and Shell Oil. The workers claim that a pesticide, dibromochloropropane (DBCP), caused sterility in men. Many of the cases were filed in the state of Texas, but judges there sent all but 4,300 cases back to Latin American courts. Costa Rican banana workers took their case directly to President Clinton when he visited that country in May of 1997 for a summit conference with Central American countries. Their plea to the President was directed toward getting U. S. judges to permit suits to be filed in U. S. courts in conformity with international law.[292]

The problem evolved in the 1960s when DBCP was a "miracle" pesticide to combat nematodes, microscopic worms that feed on the roots of banana plants. In 1977 production of the chemical was suspended when evidence was found that it caused sterility.

Two years later the product was banned by the U. S. Environmental Protection Agency. Chiquita says it stopped using DBCP about a year and a half before the EPA banned it, a contention the workers dispute. In the macho Latin American culture where a male is often judged by the number of children he fathers, sterility can be devastating.[293] Despite the DBCP cases, bananas are relatively clean when they get to customers. They rank well in federal checks of pesticide residue on fruit. At the time, Chiquita said pesticide on its bananas was ten to twenty times lower than the tolerance standard levels set by EPA.[294]

If sterility is a problem for some banana workers, it is preferable to death, the prospect for many banana workers in the Uraba region of Colombia. In 1997 workers there were caught in the midst of a life and death struggle among three forces whose aims and methods were often indistinguishable: a left-wing guerrilla insurgency; right-wing paramilitary groups; and a floundering national government with a large military and police apparatus. Along with the millions of dollars in contraband arms that come in and the drugs that go out through the Uraba region, the area also netted $400 million from banana sales in 1996. It is no wonder that everyone wants to control the region.

Colombian paramilitary groups accuse people of collaboration with the guerrillas and vice versa. The result is often murder for alleged involvement with political, labor or social causes. One human rights advocate estimated a death toll of nearly two thousand lives in Uraba in one eighteen month period in 1995-1996. The banana workers in Uraba contend not only with hard labor, low wages and pesticide but the prospect of political assassination. In the late 1990s, most of the banana workers were aligned with the Hope, Peace and Freedom Party which is known as the Hopefuls.[295]

In 2002 chaos in Columbia was still more the rule than the exception. In May 2002 the presidential vote was won by Alvaro Uribe, a hard-line conservative who favored a military crackdown on the rebels. In retaliation the country's largest guerrilla group, the Revolutionary Armed Forces of Colombia, or FARC, warned hundreds of mayors and city officials throughout the country that they would be considered military targets unless they resigned. The threat caused many officials to resign or move to secure locations causing some Colombians to joke that the country was being run by remote control. Some officials doggedly remained at their posts. Such chaotic conditions are not good for growing bananas or even the normal routine of everyday life.

The banana wars of the 1990s were not limited to Latin America. In 1994 a banana war evolved out of a civil war in Somalia, the country where the United Nations intervened in 1993 to end a famine caused largely by armed conflict between rival "warlords" seeking to control the country. After nearly two years and $3 billion in costs the U.S.-led United Nations effort failed to bring disarmament although much of the famine was relieved. The U. S. forces engaged in a futile effort to hunt down warlord Mohammed Farah Aidid. After eighteen American soldiers were killed by Aidid's gunmen and some of the corpses dragged through the streets of Mogadishu, the country's capital, the Americans and most European troops pulled out in October, 1993. That set the stage for the banana war.

Bananas are big business in Somalia and have been since Italy, Somalia's colonizer, started an export trade in the fruit in the 1930s worth tens of millions of dollars each year. When civil conflict between Aidid and his rival, Ali Mahdi Mohammed broke out in 1991, the banana industry went into decline. In 1993, Somalfruit Co., an Italian firm, pulled out altogether. The U. S. ambassador to Somalia and U.N. officials encouraged the Dole Food Company to come in. They did, hiring an Aidid clan member as its agent and staking farmers to fertilizer and supplies. By mid-1994 Dole was exporting three shiploads of bananas each month. In September 1994 the Italian firm returned to recover their twenty year investments in plantations. Within months Toyota pickup trucks with mounted antiaircraft guns manned by armed militiamen, the private armies of Dole and Somalfruit, were guarding banana plantations. The struggle threatened to escalate with several assassinations, attacks on banana truck convoys and a mortar attack on a Somalfruit vessel in Mogadishu harbor. Both companies denied any responsibility for the violence.[296] While the phrase "banana republics" may cause snickers, when politics turns to military force, the funny fruit doesn't seem to be quite so humorous.

CHAPTER 6

Selling Bananas

"Taste the better flavor that ripeness adds and you'll always want the ripeness that makes bananas so easy to digest."
Fruit Dispatch Company ad in
Good Housekeeping, Jan. 30, 1926.

The market for bananas is big and growing both in the United States and internationally. The banana is the fourth most important staple food crop produced in the world. About sixty-five million tons were produced in 1996, but because most bananas are consumed where they are grown, the world trade in bananas accounts for only about fifteen percent of production. Nevertheless, bananas are the most important tropical fruit and the fifth most important food commodity ranking behind cereals, sugar, coffee and cocoa in world trade. The United Nations Food and Agricultural Organization estimated banana exports of well over twelve million tons with a value of over five billion dollars in 1999. The world banana market is dominated by a handful of very competitive multinational corporations.[297]

Bananas have an "approval rating" that most Presidents of the United States would die for. A consumer research study has shown that fifty-eight percent of respondents had a strong liking for bananas and twenty-four percent said they "love bananas," an eighty-two percent approval![298] Thus it is no surprise that the banana is the most popular fresh fruit and king of the United States fruit market. The U. S. Department of Agriculture projected that bananas and other tropical fruit will be the leading source of increased domestic fruit consumption between 1998 and 2007. Other fresh non-citrus fruit such as apples, grapes, pears and peaches are expected to increase less than one percent annually. Fresh citrus growth is expected to remain flat in the same time period.[299]

Even in increasingly crowded produce departments, the banana is a stand-out. In the 1960s a visit to the supermarket produce department provided a choice of about sixty-five fruits and veggies. Today the average produce department sells about 275 items. Along with the old staples of lettuce, potatoes, bananas, apples and oranges, you are likely to find exotic fruits and vegetables that most Americans ten years ago did not know existed.[300] Despite the competition, bananas are the most frequently purchased produce item; sixty-three percent of shoppers purchase bananas at least once a week. For apples and oranges the figures were thirty-four percent and twenty-five percent respectively.[301] From 1988 to 1996 sales of all produce increased thirty-nine percent, but banana sales zoomed seventy-nine percent.[302]

Like consumers, food retailers appreciate the banana. According to Sherrie Terry, Chiquita's Vice-President for Marketing, "bananas are the single largest seller in the grocery store." Bananas are ten percent of all produce sales and represent one percent of total store sales. They generate one and one half percent of a store's total profit.[303] In this case the banana isn't funny; it's serious business.

Before you can gently place a bunch of bananas in your shopping cart, the fruit must be transported from its tropical home to your local food emporium. Considering the banana's delicate condition—it's much more sensitive to bruising and temperature changes than a person—this is no easy task. The banana is extremely perishable. Its rate of metabolism—remember it is a living thing—is much higher than other fruits. Most of the fruit is consumed within three weeks of harvest. Technological improvements over more than one hundred years almost ensure the delivery of a quality product although roughneck produce jockeys can still upset the apple cart. Make that the banana cart.

Since we have previously described the life of the banana through the time of harvest, we pick up the story with moving the stem of bananas out of the fields. In the early days of the industry coordinating the cutting of the fruit at the optimum time to move it to the coast to meet an arriving ship usually meant sending a telegraph message as far into Central America as possible and then having a Carib indian furiously paddle a canoe to remote farms along the coast to carry the order to managers to cut fruit. By the early twentieth century the radio made the operation much more efficient.

The cut stems, many weighing up to one hundred pounds, were loaded onto carts or donkeys for transport to rail lines which carried the fruit to the port. In some earlier operations where there was no deep water wharf the fruit had to be loaded onto small lighters to be carried out to the waiting ship. Eventually as docks were built, the amount of handling was reduced.

For many decades wagons were the accepted means of local banana transport. (Courtesy of Library of Congress, LC-USZ62-105171)

Until the 1960s when boxes were introduced, bananas were shipped to the United States, Europe and elsewhere by the complete stem. Thus they may have also transported some uninvited guests which gave rise to this request to the United Fruit Company in 1953 from a young man in Springfield, Pennsylvania. "I am very interested in spiders and snakes and if any tarantulas or a baby boa constrictor would come in on a banana boat would you please send me one and reverse the postage."[304] The company replied, "Sorry."

It is unlikely that any spiders or snakes could make the trip today. Still encased in the plastic bag that was placed over the fruit while it was growing on the plant, the freshly cut stem is most likely hung on a moving cable system which transports it to a washing and boxing station. The stems are graded for quality and the clusters are cut from the stem. Any stowaways that haven't appeared by this time are likely to drown as the clusters are placed in a water bath where they are trimmed and defective fingers are removed. A second bath lowers the fruit's temperature and stops the flow of latex which could cause unattractive stains.

The banana companies have traditionally sought ways to reduce the scarring and bruising associated with handling bananas. In 1954 polyethylene bags were introduced. An entire stem was encased in a bag after the stem was washed and brought from the fields to the railhead. At the time it was estimated that the bags saved the importer about forty-four cents per stem.[305] Plastic packaging for hands of bananas was also introduced the same year. The American Viscose Corp. supplied M. Levin and Company, a banana distributor in Philadelphia, with the first bags for retail stores. Each bag, complete with cooking suggestions, held two to three pounds of bananas.[306] Today only some hands of bananas destined for the Boston, New York and New Jersey area are packed in retail plastic bags.

A major improvement in handling came with the introduction of the banana box. Although United Fruit experimented with a box in 1925,[307] it wasn't until the industry began to change production from the Gros Michel banana to the Giant Cavendish variety that the box became the standard. The disease ridden "Big Mike" was tough in shipment whereas its Cavendish replacement was easily bruised. The Standard Fruit and Steamship Co. (now

Bananas were weighed and "banded" in this post-World War II "merry-go-round" operation. (Courtesy of Library of Congress, LC-USZ62-105174)

Dole) made a test shipment of boxed fruit in 1959. The box, primarily the work of Standard engineer B. C. D'Antoni, held forty pounds of bananas and was sturdy lightweight cardboard with air holes and built-in handles. Because the box magnified the heat and the ripening process, the water bath helped reduce the field heat before packing. The boxes were a success, and by 1963 both Standard and United Fruit were using them.[308] Boxing also meant that labor intensive cutting and packaging in the consuming countries was replaced by cheaper labor intensive work in the producing countries.[309] Today an average of sixteen hands of bananas call the box home from the moment that the two piece container is closed at the field station until it is opened at your local supermarket.

Over the years the search for the perfect box has been relentless. When United Fruit first decided to use boxes, the company went through forty box designs.[310] There have been changes in the size, number, shape and location of the holes to improve the banana's respiration en route to the consumer. Chiquita tried a one piece box with lid attached but abandoned the idea because of trade resistance.[311]

A returnable plastic container was introduced in 2001. Among its appealing features is a banana-friendly smooth, scalloped interior designed to preserve the banana's fragile touch points. The new container also requires less labor since the bananas would not have to be removed from the box at the supermarket.[312] The jury is still out on whether the new container will be adopted.

Whether by stem or by box, the banana's journey includes an ocean voyage. In the early days of the trade bananas were loaded into a ship's hold by human effort. In 1903 the first machine unloading of a ship took place in Mobile and then in New Orleans. The device used was an endless chain that dipped canvas buckets into the ship hold to lift twenty-five hundred stems per hour to the surface. Within three years better carriers doubled the capacity.[313] By 1916 horizontal conveyors were used to transfer the stems from the canvas buckets to waiting railroad cars. In the most mechanized ports, such as New Orleans, the conveyor gave way to the "curveyor" by 1948. This device developed by United Fruit carried the bananas up and down hill and around corners in one continuous operation that reduced transfers by hand.[314]

The epitome of unloading operations by mid-century came in 1952 when United Fruit opened two state-of-the-art banana terminals, one in Weehawken, New Jersey across the Hudson River from Manhattan and the other in New Orleans. Each terminal cost $1 million to construct and could unload eight thousand stems per hour. A ship carrying fifty thousand stems that previously took two days to unload required only one day. In Weehawken the latest "curveyor" equipment transported stems from the ship's hold to seventy-two railroad reefer cars and forty-five trucks all in a heated building that kept the temperature-sensitive bananas from suffering an icy winter chill. The unions were upset because the new facility required only three hundred workers rather than the five hundred who had worked at Pier 7 across the river in New York. The slicing of the first banana instead of the traditional ribbon cutting ceremony was delayed for two months until the labor issues were resolved.[315]

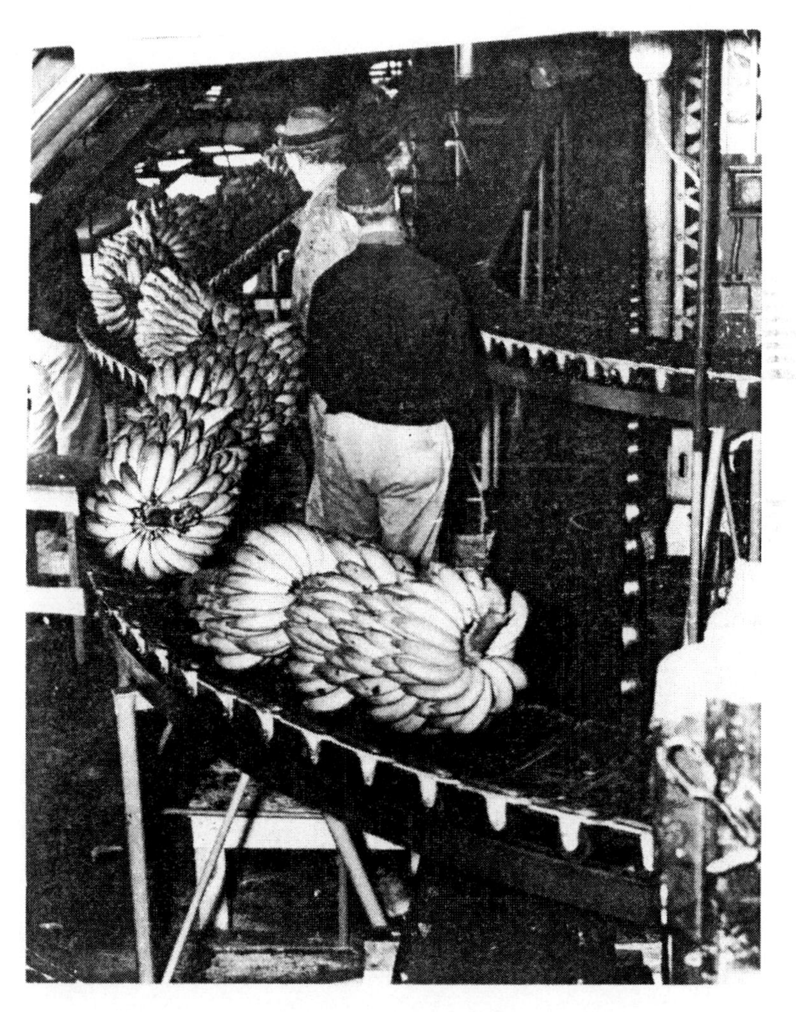

The "curveyor" was first installed by United Fruit Company in New Orleans in 1947. (Courtesy of Library of Congress, LC-USZ62-105173)

Just as Ellis Island no longer receives immigrants, the banana's point of entry has changed significantly in the past century. Undoubtedly, some of the earliest banana arrivals in the nineteenth century came through New Orleans, but when the banana became an American staple, New York City became the import leader. Until the late 1980s bananas ranked among the top three New York imports along with alcoholic beverages and cars. Today the "Big Apple" is no longer the top banana. Only two companies call at the Port of New York and New Jersey today. Bonita unloads about one ship or 185,000 boxes per week and a newcomer, Banana Distributors, Inc., planned to ship in seven million boxes annually.[316] New Orleans reclaimed the banana import title in 1953 only to lose out to other Gulf ports in recent decades. In 1948 United Fruit claimed that the number of bananas it imported through New Orleans alone in one year would reach to the moon.

From 1908-1982 Baltimore was the third largest banana port of entry. City residents ate more bananas per capita than any United States city, and politicians even used the banana as a campaign symbol. Today Baltimore sees no banana boats having lost much of the traffic to its northern neighbor Wilmington, Delaware which was numero uno in the early 1990s.[317] But it's tough to stay on top. In early 1997 Gulfport, Mississippi nosed out Wilmington by one percentage point to become the new U. S. leader in banana imports. Dole, Chiquita and Turbana moved about eight hundred thousand tons of bananas through Gulfport in 1997. Other southern ports in 1998 included Galveston, Texas; Port Everglades and Port Manantee (Bradenton), Florida. On the east coast Camden, New Jersey and Bridgeport, Connecticut also welcomed bananas. Westerners received bananas in 1998 through Long Beach and Port Hueneme (near Los Angeles), California.

At the end of the nineteenth century five port cities received ninety-five percent of all the bananas shipped to the United States. Even though the five cities represented only sixteen percent of the buying public in the United States, torrents of bananas congested port city markets, and more than eighty percent of the fruit was consumed there.[318] Residents of the hinterlands saw few bananas.

The creation in late 1898 of the Fruit Dispatch Company by Andrew Preston, one of the founders of United Fruit a few months later, was partly designed to change this situation. The railroads were to be the means of banana salvation for those deprived of the tropical fruit.

Before 1885 the rail cars that were used to transport bananas were naturally ventilated with iron bars in side windows and doors. It quickly became clear that the delicate banana needed protection from heat and cold. An irate merchant sent the Boston Fruit company two two-cent postage stamps in payment for two railroad cars of spoiled bananas. "That was all they were worth," he said.[319] The solution was the refrigerator car that could be iced in warm weather and heated in cold. The banana spread everywhere the rail network carried refrigerator cars. They became so ubiquitous that Don Herold, a popular columnist for the *Indianapolis Star*, wrote that, "Bananas grow in boxcars and are very easy to pick."[320]

Along with the refrigerator car, the banana got a companion known as a Perishable Freight Inspector or "messenger" of the Icing and Messenger Service. A typical journey of perhaps 115 carloads of bananas might begin in New Orleans. After the reefer car was loaded, it was moved to a scale where it had been previously weighed while empty. The difference in the two weights was the weight of the bananas. The car was then moved to the icing dock to load ice, adjust ventilation hatches or to place heaters in cooler weather. The "messenger" spent most of his time in the caboose while the train headed north on railroads such as the Illinois Central. His job at stops along the way was to sample the temperature of the banana pulp at the top and the bottom of the stem by inserting a thermometer into banana fingers. Cooling, heating, and ventilation in the cars was adjusted accordingly. Most Perishable Freight Inspectors worked in stationary locations such as Fulton, Kentucky, a key stop in the Mississippi Valley. In cold winter weather entire trains could be moved into a building called a "warm house." The United Fruit Company alone had about two hundred inspectors to pamper the very important "passengers."[321]

"Fruit Growers Express Company" icing station in Atlanta, Georgia in 1937. (Courtesy Library of Congress, LC-USZ62-116753)

When the cars of bananas from the Fruit Dispatch Company arrived, distributors/jobbers found instruction cards inside the car doors. Printed in both English and Italian, the latter an indication of the ethnic background of many in the banana business around 1917, the card gave instructions on how to protect the bananas against cold weather including using canvas to prevent a cold wind from blowing against the bananas. It advised to use plenty of hay to keep the bananas warm.[322]

The constitution of the banana hasn't changed, but transportation methods have changed drastically. The ice, hay and canvas have given way to containers with controlled atmospheres. The typical procedure is to stack the forty-pound boxes on pallets which are then loaded into a container on a railroad car at the tropical farm location where the boxes were packed. Each container may hold as many as 970 boxes. The containers are then moved to the

Complete stems of bananas being prepared for shipment from the distributor to the retailer in the mid-1940s before the 1960s introduction of the cardboard box for hands of bananas. (Courtesy Library of Congress, Prints and Photographs Division)

port area where they are attached to a controlled atmosphere system that will keep the temperature between 57°F and 58°F during the shipping process. In 1993 Chiquita also began to use a controlled atmosphere process that slows down the fruit's respiratory process by adding high-purity nitrogen and lowering the oxygen level to less than ten percent. In a sense the rock-hard green bananas are put to sleep. This system allows for more flexibility in the ripening process when the fruit arrives at ripening rooms in North America or elsewhere.[323]

Both Dole and Chiquita operate large fleets of refrigerated container vessels. Wal-Mart and Dole occupied the top two positions in the volume of all containerized U.S. imports in 2002 while Chiquita ranked sixth. Chiquita's "Great White Fleet," originally so named because the ships were painted white to reflect the heat of the tropical sun, consists of twenty vessels either owned, operated or chartered by Chiquita. In 1995 the company owned some four thousand forty-three foot containers and about twenty-five hundred chassis to transport them.[324] Although in the mid-1990s Chiquita contemplated using an intermodal transport system including railroads, virtually all bananas today are moved by truck to the ripening rooms of distributors/jobbers or supermarket chains, their last stop before the local market.

A modern container ship of Chiquita's "Great White Fleet"
(Courtesy Chiquita Brands International, Inc.)

Getting bananas to market with the proper degree of ripeness is no easy feat. In ancient times the Chinese knew that pears could be ripened by exposure to the smoke of incense burned in closed rooms. Many years ago California and Florida oranges were "degreened" by exposure to fumes from kerosene stoves or exhaust from a gasoline engine in special coloring rooms. The active agent in the ripening process was the ethylene in the fumes.[325] Ethylene is given off naturally by ripening fruit, and temperature controls the rate of ripening. In the early days of the industry bananas were ripened in cellars heated only by open gas flames. Experienced men were supposed to sense or feel the temperature of the room. In hot weather the fruit could "cook" causing losses to the dealer. Because jobbers had no satisfactory means of regulating the ripening process, the market was often either oversupplied or bare of bananas.

Between 1908 and 1915 some jobbers began to experiment with refrigeration in their ripening rooms. The United Fruit Company constructed two experimental ripening rooms in New York City and began scientific and practical studies of the ripening process in the early 1920s. By 1928 the company had established a dealer and jobber service department to advise on the handling of the fruit. By mid-century eighty percent of the jobbers' ripening rooms were equipped with automatic systems where, as the *New York Times* observed, the temperature and humidity were as carefully controlled as the sprinklers on the greens of a fine golf course. A steady supply of bananas year round was assured.[326]

The banana companies were positively obsessive in educating jobbers and dealers on the proper handling of bananas. United Fruit's marketing arm, The Fruit Dispatch Company, used its publication *Fruit Dispatch* to inform its industry readers of the latest developments in ripening room technology, banana baskets or containers, and proper handling. In 1917 the publication admonished store clerks who pulled fingers from the stem resulting in partially unpeeled bananas. A special banana knife was offered to retailers for forty cents so that fingers could be *cut* from the stem. The magazine warned retailers that bananas torn from the stalk are unsanitary, and lectured that "nature has sealed the banana in a germ-proof package; do not let it be broken."[327]

The treatment of a banana in a modern ripening room is not

unlike that of a patient in a hospital intensive care unit except that the banana should be in good condition when it arrives. The ripening rooms are operated by banana wholesalers or by large supermarket chains that do their own ripening. If all has gone well in its journey from the tropics, the banana will have been untouched by human hands since it was packed in its box. It should arrive with a pulp temperature of 58° to 60°F. It is time to wake up the bananas, and the goal is to ripen them to color level #4, more yellow than green, during a four to eight day ripening cycle. The boxes, stacked on pallets in such a way as to permit uniform air circulation, are placed in a ripening room which is then sealed so that the bananas can be gassed with ethylene. Bananas themselves produce ethylene, but small amounts of ethylene (one cubic foot per 1000 feet of ripening space) are added to trigger a uniform ripening process. Excessively high concentrations of gas could ignite and blow the fruit to banana heaven such as happened in Pittsburgh in 1936. The little domes on the towers of St. Stanislav's Catholic Church in the "Strip District" may not be heaven, but they were blown away when a gas explosion tore through the Pittsburgh Banana Company building across the street.[328]

A high relative humidity level of 85 to 95 percent is maintained for good ripening, and a constant monitoring of temperature is required because bananas produce considerable heat during ripening. This is known as the heat of respiration. The rate of heat production increases to a peak called the climacteric about half way through the process, and the fruit continues to color after this peak. When color level #4 is attained, the pulp is brought to about 58°F, and the fruit is ready to ship to the store. The usual time for the complete journey from harvest to the supermarket is from five to eleven days.

Weather conditions during shipment to stores will affect the quality of the fruit. In warm weather pulp temperatures should be 57° to 58°F prior to shipment. In cold weather 60° to 62°F is the recommended range.[329] If the bananas suffer a chill by exposure to temperatures below 56°F at any time after they have been cut from the plant in the field, they may suffer injury leading to a dull color, poor flavor or a failure to ripen.[330] No one benefits from a banana that has suffered frostbite in a Minnesota winter.

By 2002 a new state-of-the-art temperature control system for trailers was being introduced that would permit the ripening of bananas while still in transit. The "SmartAir" system developed by Carrier Transicold of Syracuse, NY creates a processing plant on wheels similar to land-based banana ripening rooms. Shippers can have greater control over the vital temperature of the bananas, especially over long distances.[331]

Over the course of more than a century the goal of getting a quality banana to the supermarket has been pursued with science, technology, ingenuity and organization with almost military-like precision. At the same time the goal of getting the consumer to buy bananas has also been pursued with determination and creativity.

In some produce departments today you might come across a kiwano, also known as a horned melon. It has a bright orange shell covered with spikes. Frieda's, a Los Angeles wholesale distributor of unusual produce, calls the kiwano a "masterpiece of nature." It has a brilliant green interior that has the taste and texture of a crunchy, seed filled, mild melon. Unless you were told, you would probably not know that a kiwano is ripe when it is a bright, deep orange color. You might also not know that it should not be refrigerated. To many Americans at the beginning of the twentieth century, the banana was equally mysterious and misunderstood.

Much of the advertising effort of the banana companies for at least the first half of the century was directed to educating the public about when a banana was fully ripe, a message that still may not be understood by some consumers. The *Journal of the American Medical Association* described the essence of the problem in 1917 when it said, "The chief reason for the unfavorable reputation attained by the banana when eaten uncooked, appears to lie in the failure of most persons to understand what a ripe banana really is."[332]

Correcting wrong notions about the banana's effect on human health was directly related to the ripeness issue. For a long time the banana was regarded in some quarters as indigestible and unhealthy. Apples, oranges and cherries were usually ripe when purchased and could be eaten immediately. The banana was in an unfortunate position because it was seldom ripe when purchased. If it was eaten immediately, indigestion and a dissatisfied consumer was the result. It was a problem the banana companies had to address for decades.

Citing the nutritive values of the bananas in the early twentieth century did not carry much clout because nutrition was the new kid on the block in the science community. The novelty of nutrition started around the turn of the century with the identification of the first vitamins. The word vitamin was first used in 1911, and in 1913 Vitamin A was the first to be identified . Vitamin B6, found in moderate amount in bananas, was not identified until the mid-1920s.[333]

Despite these problems, by the time of World War I people recognized the banana as a cheap food of the masses. An editorial in the *Houston Post* in 1913 noted that the home grown apple was beyond the reach of the average consumer while the foreign grown banana had increased in quality and declined in price. The editors also pointed out that apples sometimes sold by weight for ten times the price of bananas.[334] Even in early 2002 when bananas were selling for an average of fifty-two cents per pound, Red Delicious apples cost nearly twice as much at ninety-two cents per pound.[335]

A check of grocery ads in the early twentieth century found that bananas typically sold for fifteen cents a dozen and sometimes for a penny each.[336] In 1916 the *Boston Post* was disturbed by the high cost of living and noted the outrageous rise in the prices of four staples: beans, moth balls, collars and corncob pipes. The newspaper found that the best bet for something for which the price had declined was the banana which it said was on the verge of ousting the bean and beefsteak in Boston.[337]

Consumers recognized a good thing when they saw it and thwarted an attempt to put bananas on the tariff list in 1913. While considering the Underwood-Simmons tariff bill, the Senate

Finance Committee inserted a five cent per stem duty on bananas. The legislators expected that two million dollars could be raised on the forty million stems of bananas imported each year and assumed that the "banana trust" could absorb the expense. The full Senate changed the rate to one tenth of one cent per pound, but the public would have none of it. Lobbyists and the press charged that the proposal was a tax on the "fruit of the poor man." President Wilson wanted to reduce the cost of living for the little man and was also opposed. The final tariff bill contained no duty on bananas, but apples, peaches and other fruit were subject to a tax of ten cents per bushel.[338] There has never been a United States tariff on bananas, and the yellow fruit remains a relatively low cost food.

Even with rising wartime prices, the Dallas Fruit Co. in May 1917 advertised bananas for twenty cents a dozen in an ad with the headline "Cut High Cost of Living by Eating Bananas" and went on to state "Thomas Edison, the great inventor, attributes his good health to intelligent use of the banana as his main sustenance."[339]

A wartime ad by the Fruit Dispatch Co. in 1918 covered all the bases. It proclaimed that the banana was wholesome and nutritious, delicious and easily digested, always and everywhere available, good cooked or uncooked, easy to handle and convenient for the dinner pail, and sealed by nature in a germ-proof package. Exploiting the special concerns of wartime, the ad drove home the points that the banana was cheap and the poor man's food along with the patriotic idea that it was produced without drawing on the nation's resources.[340] What more could anyone want in a food?

Home economists were trotted out to demonstrate that the banana gave people more bite for their buck than other foods. In 1917 a Hunter College home economist compared apples, oranges and bananas along with some other foods. She found that although the three fruits were comparable in average cost per pound (Four cents for apples and bananas and five cents for oranges), that after allowing for the refuse, the banana provided more calories per pound, 460, of the edible portion of the fruit. The apple and orange provided 290 and 240 calories respectively.[341] An Indiana banana distributor's ad in the same year touted similar findings by a well known dietetic (the term used at the time) printed in a magazine for women. The banana "in relation to heat producing energy, has one third greater value than the apple, and twice the value of the orange. The amount of tissue-building material is identical with that of milk. The sugar it contains is a valuable asset," said the ad.[342]

The status of the banana in America by the time of World War I was well described by an editorial in the Lexington, Kentucky *Herald* in 1917. The writer recalled that the first banana seen in Lexington was forty-five years earlier when relatives who had attended the Mardi Gras in New Orleans brought back for children a bunch of bananas that were not then available in Lexington. By 1917 1,350 stems were sold each week in Lexington, and the *Herald* could say, " . . . the banana, the Sunkist orange, chewing gum and Bull Durham have conquered the American nation."[343]

The banana business boomed along with the American economy of the 1920s, and typical of the era, advertising became a major tool to improve sales. Three quarters of all bananas from Central America and the Caribbean were shipped to the United States. The United Fruit Company alone accounted for about half of the imports. Standard Fruit, the Cuyamel Company and Mexican growers provided the rest. Honduras was the number one producing country.[344]

The banana companies experimented with brand names as a way to differentiate their produce from that of competitors. At the time of World War I, Fruit Dispatch, the marketer for United Fruit, sold their bananas under the name "The Best." This was changed to "Unifruitco" bananas, and in 1926 it became "Unifruit Bananas: A United Fruit Company Product." In 1928 United used triangular stickers with the label "Unifruit" on their bananas. Standard Fruit used the label "Frisco" for its fruit shipped in the 1920s.

An indication of an old marketing problem that wouldn't go away was the slogan United Fruit was using in 1922, "Eat More Bananas, But They Must Be Ripe."

A writer for the *Detroit News* in 1926 probably did more harm than good for the cause. Citing three consulting chemists, he offered that " . . . the banana slips through the interstices of the consumer with ease and celerity, playing around in the gastric juices like a kitten in a bowl of cream."[345] Such an endorsement must have been an immeasurable help for sales.

Along with the digestibility issue was opposition in some quarters to bananas as food for small children. The Fruit Dispatch Co. thought the problem particularly acute with small town doctors who looked with disfavor on the yellow fruit. A 1926 article in *Good Housekeeping* offered little support. The story provided detailed menus suggested for children at ages one, three, six and ten. It proposed that toddlers in their second year should have orange juice, stale bread crust (for tooth exercise) and strained prune pulp in their diet. By six years the child had graduated to stewed prunes and sliced oranges. The ten year old diet included round steak, baked apples and stewed figs. Bananas were no where to be found in the recommendations.[346]

The prohibition on bananas for young children was deeply ingrained as Dr. Benjamin Spock, the famous baby doctor, made clear in his autobiography *Spock on Spock.*

> Another friend was Chunky Robbins. He was a rather bossy, slightly fat boy, very cheerful and friendly. When I was eleven years of age I was playing in Chunky's backyard one morning with five or six other boys. Chunky went into the house for some reason or other and came out holding a big bunch of bananas and proclaimed, "Everybody's gotta eat a banana!" Well, my mother, who followed exactly the health precepts laid down by Henry Holt in *The Care and Feeding of Children*, had been led to believe that bananas were dangerous food for young children. She said that we couldn't even taste a banana until we were twelve, and then we could only eat half a banana. So I was confronted with a serious crisis, but it only took me a few seconds to realize I was much more scared of my mother than of Chunky Robbins, so I

> said in a timid voice, "My mother says I can't have half a banana until I'm twelve."
> A year later, when I first had permission to eat half a banana, I was rather apprehensive, imagining that if I was more sensitive than the average child it still might kill me.[347]

By the mid-twenties the United Fruit Company was the dominant banana player and had established a year round distribution network through its Fruit Dispatch Company that had forty-nine branches in the United States and Canada. Although the sale of bananas was increasing, sales were not growing as fast as the population. Along with the health issue, an aggressive citrus fruit campaign by the California Fruit Growers' Exchange and Florida Citrus Exchange convinced the banana company that aggressive advertising was needed.

The Fruit Dispatch Company launched a massive campaign in 1925 designed to reach one-fifth of the country's population living in five Mid-Western states. The campaign promoted the value of the banana as a body builder. "Yes! Bananas—The Body Builder" was the prominent slogan in newspaper ads, billboards and counter cards for retailers. Every dealer got a thirty-two page manual with the campaign's marching orders to convince the public of "the undeniable supremacy of the banana over any other fruit on the market today." Each dealer got a color chart showing bananas in four stages of ripeness to educate consumers when the banana was best for eating. Different weekly newspaper ads ran for fifty weeks and showed people in various walks of life consuming bananas. "The He-Man Food," was the title of one ad which proclaimed, "Healthy, husky fellows whose muscles turn the cogs of commerce need food that fills and food that builds." Another ad stressed the fact that bananas contain no acid. Another highlighting a small boy declared, "Let Him Eat All He Wants." If the 3,446,515 newspaper readers did not get the message from the ads or the 1,245 billboards, some illuminated at night, they could get a free booklet, *100 Ways to Enjoy Bananas,* from their grocery store.[348]

In 1926 Fruit Dispatch inaugurated the first national advertising campaign for bananas. The first objective of the campaign was to counteract the adverse health propaganda, and the second goal was to increase the market for bananas by creating new uses for the fruit. Four color ads ran in the *Saturday Evening Post* and four women's magazines. The art work was exquisite, and it was the first time that this type of color photography ever appeared in commercial work.[349]

A 1926 United Fruit Company magazine advertisement to educate consumers about banana ripeness. (author's collection)

The banana ads in magazines such as *Good Housekeeping* had plenty of competition. The Hawaiian Pineapple Canners lauded canned pineapple as better than fresh. There were ads for oranges, cranberries, canned asparagus, California walnuts and Sunkist Lemons. Less appetizing were ads for "FLY-TOX" anti-bug spray, Hoover "positive agitation" vacuum cleaners, and Kleenex "absorbent kerchiefs."[350]

It was expected that homemakers would learn about banana ripeness and tasty dishes made with the yellow fruit along with reading stories such as "Wedding Bells, the heart-to-heart story of a girl who found that wedding bells ring in tune only if you hear them with the right man," and "The Brute, a story of a man whom everybody loved except his wife." Perhaps after reading "Our Second Line of Defense, interviews with prominent educators throughout the country to find a campaign plan that will keep our young folks marching forward, instead of falling back," mothers would clip the coupon in the banana ads to send for a booklet entitled *From the Tropics to Your Table.* The eighty recipes for new and attractive ways of serving bananas were prepared by the former chef to His Majesty, Albert, King of the Belgians.[351] Of course the booklet used a color chart to advise when bananas were ripe and warned not to put them in the refrigerator. By May of 1926 Fruit Dispatch had printed one million copies of the recipe book and was receiving one thousand requests per day.

Not to be outdone by industry leader United Fruit, in 1926 Standard and Cuyamel together created the Banana Distributing Company to compete with the United Fruit distribution network which serviced eighty to ninety percent of the United States market. The new distributing company also issued a recipe book titled *Everyday Banana Recipes.* The two companies then agreed to work with United to energize banana jobbers and launch an intensive advertising campaign.[352]

The companies left no stone unturned. Cooperative ads were arranged with the Minute Tapioca Company for bananas with gelatin. An ad with the Kellogg Company for bananas served with "All-Bran" discretely stated that ripe bananas can be good for people with diets that need watching or correcting. *Fruit Dispatch* trumpeted the efforts of its salesmen and jobbers. Because there were between 125 and 150 thousand soda fountains in the United States in 1926, jobbers were urged to push bananas for banana splits and counter displays, but again with the caution that the "bananas must be ripe." In April 1929 the 530 Liggett Drug Store soda fountains from Maine to California featured twenty-cent banana splits and sold 352,789 of them worth $70,000. They expected to sell 750,000 splits in a June promotion.[353]

The summer of 1926 was particularly fruitful. During the first week of May the banana on a stick was introduced at a street carnival in Philadelphia. A ripe banana, with peel removed and a stick inserted, was covered with chocolate. It sold for five cents and demand exceeded the supply. The week of July 26–31 was declared national banana week. In August *Fruit Dispatch* reported that the Hoefler Ice Cream Company of Buffalo, New York had developed a new ice cream which in "appearance resembles a lovely serving of watermelon." It consisted of banana, strawberry, chocolate and other ice creams and bore the name "Banana and Pickaninny Ice Cream."[354]

Banana enthusiasm extended beyond the soda fountain. The Fruit Dispatch Company encouraged jobbers to sell bananas to roadside stands and gasoline stations. "If there are two attendants at a gasoline station which is carrying bananas as a side line, one can take care of the car while the other makes a selling talk for bananas."[355] And we think we have convenience stores!

Still fighting the problem of the public's lack of knowledge of banana ripeness, the message was taken to the schools. In 1926-27 Fruit Dispatch sent 110,000 charts on the growth and cultivation of bananas to schools. A year later a special edition of the three stage ripening reference chart was sent to teachers along with the advice to instruct the students in when bananas were ripe. In a 1929 publicity campaign the company used the *Journal of the National Education Association* to offer teachers booklets on the food value of bananas along with lessons on the fruit and the countries where they grow. The ever present ripening charts were part of the package.[356] A merchant in Cairo, Illinois seized the occasion of a school teachers' convention to hang several bunches of ripe bananas outside his store with a sign inviting all "school marms" to help themselves.[357]

In 1929 as the country was on the verge of a stock market crash and the Great Depression, the Fruit Dispatch Company was gearing up to conduct research on all matters affecting growing, transporting, handling and ripening of bananas. The company conducted a survey to learn more about banana consumer demographics. Interviews were done with 8,500 consumers, 92 jobbers, 1,741 retailers and 341 institutions (tea rooms, hospitals, restaurants, boarding houses, soda fountains) in Massachusetts, Georgia, Ohio and Iowa. People of all income levels living in cities, suburbs and rural areas were questioned. The findings demonstrated a need to do some (surprise!) health education and a need to win favor among the "well-to-do." Over ninety percent of the consumers reported that they liked bananas, but only about one-half regularly ate them. The banana was perceived as a "poor man's food." Among those who never ate bananas, a majority thought the fruit indigestible. Others believed that bananas were fattening. The survey surprisingly showed that people tended to stop eating bananas when they reached the age of forty to fifty.[358] United Fruit concluded, "Bananas do not suffer because they are unknown. Rather do they suffer because they are *incorrectly known* [italics in original]. They suffer too from lack of style appeal."[359]

Along with the Standard Fruit and Steamship Company and the Di Giorgio Fruit Corporation, Fruit Dispatch created the Banana Growers Association to create a unified banana ad campaign directed to women. Among the magazines selected, *Ladies' Home Journal* was chosen because it was a circulation leader. *Good Housekeeping* was thought to be of unusual class. The *Woman's Home Companion* was selected because it was read "by more women who actually do their own house work," and *True Story* provided "further coverage of the wage-earners." Seven other magazines, including two juvenile publications, completed the package. The Association figured that each promotion dollar bought magazine access to 275 "good" homes for their instructive message.[360]

As 1931 dawned, a newly reorganized Sales Promotion Department at Fruit Dispatch used United Fruit's company magazine, *Unifruitco,* to issue the marching orders—they were called the "13 Banana Steps"— to make the 1931 advertising campaign a success. The advertising messages, it was noted, "point out the food value of this popular fruit, keeping folks right in step with the moving march of food events—the vitamin discoveries—the latest, tastiest recipes." Children would learn from "our clever banana game and cutouts" offered in juvenile magazines. The "educating classes [doctors, domestic science workers, teachers]" will "teach the masses." The latest banana recipe book, *The New Banana,* was being requested by thirty thousand persons a month. Each employee was asked to send in the names and addresses of six friends who would welcome the book. Finally, it was reported that " . . . highly trained forces of men are moving about the country under orders" to see that jobbers and retailers were handling the fruit in the best possible way.[361] The campaign was nothing less than mobilization for a new deal for bananas two years before Roosevelt's "New Deal."

The Great Depression made marketing bananas more difficult, but World War II made it almost impossible. Banana imports of 1,825,000 tons in 1928 dropped to 550,000 tons by the depression year 1933. The retail price of 9.9 cents in 1928 declined to 6.5 cents in 1932.[362] In the war years from 1942 through most of 1944 bananas were virtually unobtainable in most of the United States. The reason for the banana dearth was simple, lack of transportation. Even so in 1943 the United States received five-sixths of total world exports.[363] Because virtually all of its refrigerated ships had been requisitioned by the government, United Fruit had to make do with eleven old naturally ventilated vessels some of which had been built in the 1890s. This fleet had to operate through Tampa, Florida because almost every other port was closed to regular commercial traffic.[364] In 1943 United Fruit imported only twenty-five percent of its available supply, and in the first months of 1944 it had to destroy sixty percent of its crop for lack of transport.[365]

Newspapers recorded the first post-war shipments with fan

A 1931 United Fruit Company magazine advertisement that offered the company's "Children's Game." (Courtesy of Chiquita Brands International, Inc.)

fare. In May 1946 United's ship, *Junior*, arrived in New Orleans with a shipment of eighty thousand stems, a port record. The first load of United Fruit bananas for New York City, twenty thousand stems from Guatemala, was unloaded on Dec. 3, 1948.[366]

The banana-starved Europeans welcomed post-war bananas.

In 1948 the Viennese got their first bananas in six years as a result of a barter deal between the Italians and the Russian occupation forces in Austria. A Soviet officer inspected the shipment. He had never seen a banana before, and he chomped a big bite—skin and all. It tasted terrible. His ruling: "Unfit for Russian military personnel, dispose of them on the Austrian economy."[367]

Supplies were one thing; prices were another. By September 1946 oranges and bananas were the only fruits or vegetables still under Office of Price Administration (OPA) regulation. Ceiling prices were set weekly. For the week of September 5 the ceiling for California oranges was fourteen and one-half cents per pound. Central American bananas were twelve cents per pound and all other bananas were eleven cents per pound. Several months later Samuel Zemurray, president of United Fruit, claimed that retail prices, even allowing for rail freight and handling, were too high compared to the price, five and one-half cents per pound, charged by United Fruit at Southern ports. That price was considerably less than the price charged at public auction before price controls. Zemurray suggested that consumers could refuse to buy and retailers should reduce prices to a more reasonable level.[368]

The long suffering consumer sometimes found the banana supply disrupted by longshoremen strikes and other labor disputes between 1946 and 1951. In 1946 the patients of New York City hospitals feasted on six thousand stems of bananas that were kept out of normal distribution channels by a truckers' strike. The bananas were taken by tugboat from the ship in the harbor and then moved by city-owned trucks. In 1948 a strike by longshoremen prevented several ships from unloading in New York, and the strike threatened to spread to Gulf ports when the ships were diverted there. During a 1951 strike, more than a half-million dollars worth of bananas rotted on docked ships in New York.[369]

Bananas with black peels, but not rotten, is what you get if you put bananas in the refrigerator. For decades the banana companies waged a campaign to prevent such a faux pas. The Grand Rapids Refrigerator Company ran an ad for the Leonard Cleanable Refrigerator in the January 1925 issue of *Good Housekeeping*. The

A Minnesota grocer "writes-up" the consumer's purchases, including bananas, around 1947. (Courtesy Library of Congress, Prints and Photographs Division)

ad pictured a fridge with a bunch of bananas hanging in it and quoted a customer saying, "I removed the shelves and put a string of bananas in—and, believe me, they kept, didn't even ripen." United Fruit's equipment department sent a letter of complaint to the offending advertiser.[370]

In 1944 this chilling concern resulted in the creation of a radio advertising jingle and a personality that United Fruit eventually adopted as its corporate name. As United Fruit looked toward the end of the war and the resumption of a normal banana supply, John Werner, the head of Fruit Dispatch, wanted to educate consumers on the proper care of bananas and that they should only be eaten when "flecked with brown." The Batten, Barton, Durstine & Osborn agency was given the task of developing a sixty-second radio ad. The Chiquita jingle that warned consumers not to put bananas in the refrigerator was the result. The jingle quickly became a popular song hit. The message apparently got through, at least to students at a Mid-western university that voted Miss Chiquita "the girl they'd most like to get in a refrigerator with."[371]

The illustrated Miss Chiquita character was created by artist Dik Browne who also created the Campbell Soup kids and the Hagar the Horrible cartoon. The original illustration was an animated banana, but in 1987 the illustrations became an actual woman done by artist Oscar Grillo who also created the well known Pink Panther. An animated Miss Chiquita appeared in a series of eighty-second film commercials that ran in 350 theaters in the United States in the mid-1940s. As a recognizable corporate symbol, Miss Chiquita ranks right up there with Aunt Jemima of Quaker Oats, Betty Crocker of General Mills, and Charlie the Starkist Tuna.[372]

Of course Miss Chiquita made personal appearances and probably the most famous person to play that role was Elsa Miranda, no relation to the movie star Carmen. The media had a field day in 1947 when Miss Chiquita, famous for telling the world to keep bananas out of the fridge, appeared in Baltimore at the christening of the Great White Fleet's new fully refrigerated ship S. S. *Sixaola* built to carry millions of bananas in chilled comfort.[373]

To mark the fiftieth anniversary of the character, Chiquita Brands conducted a contest to find Miss Chiquita 1994. Contestants had to be able to sing the Chiquita jingle while wearing a ten pound bowl of fruit on their heads. Selected for the honor from among 120 Miss Chiquita wannabes, including a male auto mechanic from Staten Island, was Elizabeth Testa from Syracuse, New York.[374] The new Miss Chiquita got to samba across the country on a twenty-five city "Good Will, Good Nutrition" tour.

The Chiquita success initially helped sell bananas regardless of company. The situation changed in the early 1960s when John M. Fox, United's Executive Vice-President, decided to recapture the Chiquita name for United Fruit alone by applying a Chiquita sticker to every third banana. When Fox, the "Orange Juice Man,"—he had been president of the Minute Maid Orange Juice

Company—made the sticker announcement at an executive meeting, some of the banana old-timers laughed. "One of them—a Southerner—did a quick calculation and almost collapsed in hysteria. 'Shee-it, man,' he said to Fox, 'Do you realize what that would amount to in a year? One billion stickers! You've got to be out of your mind!'" The actual requirement for the first printing was two and one half billion stickers. The printer's representative fainted when he received the order.[375]

The company called on machinery and equipment designers to determine how to apply the pressure sensitive labels. Some calculated it would cost almost ten dollars per label. A young Honduran laborer devised the solution; apply the stickers by hand. United Fruit thus became at the time the largest single user of pressure-sensitive materials in the world, and banana stickers are now collected like stamps.[376]

The sticker innovation was arguably the most important marketing concept in fresh produce history. A banana is a generic commodity; with a sticker the banana is a brand. It worked for Chiquita, the name sixty-one percent of consumers say first when asked to name a banana brand without assistance. Dole is named by thirty-two percent. Consumers prefer Chiquita over Dole, sixty-five percent to twenty-eight percent.[377] In a 1998 quality perception ranking of 270 brands across fifty product categories, Chiquita ranked third behind Kodak film and Craftsman tools and just ahead of Hallmark greeting cards. Among all food brands Chiquita was number one. In the same quality ranking Dole bananas was ranked 35, Del Monte bananas 72 and Bonita bananas 156.[378]

Not all innovations worked as well for Chiquita as branding bananas. In the 1950s the Kellogg Company introduced a new version of its famous breakfast cereal. Banana puree was molded into a banana shape, cut in slices and freeze dried. The dry product, which looked like wooden nickels, was packaged with cornflakes. The banana slices took as long to hydrate with milk as it took the flakes to become an unappetizing soggy mass.[379] Another connection with Kellogg had better results. The Corn Flake makers offered ten and one-half inch Chiquita oilcloth dolls as a premium. By 1994 the doll was a collectible worth sixty-five dollars.[380]

"Bananas are better because they have no bones!" That was just one of 345,000 "Bananagrams" sent to United Fruit by 155,800 young people in the late 1950s in a contest that invited the youngsters to write witty little anagrams. "Bananas are best—with freckles on their vest" and "Health yourself to bananas!" were two other entries that won twenty-five dollars because they were published along with company prepared illustrations. All entrants received a 45rpm record of teen-favorite Rusty Cannon singing "Bananas What a Crazy Fruit."

A 1967 Chiquita advertisement alerting buyers that a "Chiquita" banana isn't your ordinary banana, an effort to establish the "Chiquita" brand identity. (Courtesy Chiquita Brands International, Inc.)

In the 1960s some Chiquita copywriters got carried away with an ad which seemed to appeal to prurient interests. The ad which appeared under the headline "What does a banana have to be to be a Chiquita?" showed a ripe yellow banana whose length of eight inches and its thickness of one and one-quarter inches was indicated by black markings. The text of the ad read:

> It's sort of like passing the physical to become a Marine. The banana's got to be the right height. The right weight. The right everything. Right off, it has to be a good eight inches along the outer curve. And at least one and a quarter inches across the middle. It has to be plump. The peel has to fit tightly. The banana has to be sleek and firm. It has to be good enoughto get through a 15-point inspection. Not once, but three separate times. Occasionally, though, a banana comes along that's got everything going for it—except maybe it's a smidgen under minimum length. We should pull it, we know. But our inspectors have hearts,too. Which is why you may sometimes find a Chiquita Brand Banana that isn't quite eight inches long. Come to think of it, though, you sometimes find Marines named "Shorty," too.

Life magazine was not amused, and refused the ad. Other magazines were too polite to raise a question and ran it.[381] Perhaps Chiquita needed such attention grabbing ads because by the 1960s the company's long dominance of the industry was being threatened by an old rival, Standard Fruit, and some additional competitors. The business of selling bananas was changing.

CHAPTER 7

A Banana Eat Banana World

"We're one of the few companies that's been able to grow market share. This is a tough industry to grow market share in."
John Loughridge, Vice-President of Marketing, Fresh Del Monte Produce.[382]

After decades of preaching by United Fruit and other banana companies, by the 1960s most Americans had learned when a banana is ripe and knew better than to put them in the refrigerator. But the days of United Fruit's almost unbridled sway over the industry was about to change. In 1958 after a four year antitrust battle with the United States Justice Department, United Fruit signed a consent decree by which it agreed to establish within ten years a competitor at least one third its size. At the time Standard Fruit was about one third the size of United. Thus, the Justice Department envisioned an industry with at least three relatively equal competitors.

Standard Fruit and Steamship Company was incorporated in 1923 as an outgrowth of the New Orleans banana pioneer Vacarro Brothers firm. Castle & Cook, incorporated in 1894 and with major interests in Hawaii and marketing under the Dole brand name, purchased fifty-six percent of Standard's common stock in October 1964, and in December 1968 Standard became a wholly owned subsidiary of Castle & Cook. Standard used the brand name "Cabana" beginning in 1959 and added the name "Tropipac" in

1964 for its bananas from Ecuador. In 1972 the company adopted the Dole name, a brand identification long known to consumers.[383]

The rise of Standard/Dole was rapid. In 1955 Standard had about fifteen percent of the U. S. banana market; United Fruit/Chiquita's assets were about ten times those of Standard. By 1968 Standard's banana sales were in excess of one-third of the world trade in bananas, and by 1973 Standard, using the Dole label, moved ahead of United Fruit/Chiquita in United States sales for the first time in history.[384] Completing the name change game, Castle & Cook adopted the corporate name of Dole Foods in 1991.

Del Monte became the third largest banana producer in 1972 when it bought United Fruit's plantations in Guatemala. The sale by United was still part of the fall-out from the government's antitrust action. As a result of this activity, the three largest banana companies in 1972 were Standard Fruit (Dole), United Fruit (Chiquita) and Del Monte. Together they controlled ninety percent of the Central American banana trade. They had ninety percent of the United States and sixty percent of the world market in 1972.[385]

Of the three banana "biggies," Del Monte—Fresh Del Monte Produce, Inc. is the official name—is truly a transnational corporation. The company's roots were in the California packing and canning industry of the 1850s. The Del Monte name was first used in the 1890's by the Oakland Preserving Company which by 1967 officially became the Del Monte Corporation known mainly as a canner of fruits and vegetables. The company entered the fresh banana business in 1968 when it bought the West Indies Fruit Company and formed Del Monte Banana Company. It acquired United's Guatemala operations in 1972.

A dizzying string of ownership changes of the Del Monte Corporation began in 1979 when the company was acquired by tobacco giant R. J. Reynolds Industries which was then merged with Nabisco Brands in 1985 to create RJR Nabisco. In 1989 RJR Nabisco was taken over in a leveraged buyout by an investment fund controlled by Kohlber, Kravis and Roberts. Del Monte's fresh produce business, which by then included bananas, pineapples and melons, was sold to a British conglomerate, Polly Peck International, owned by a Turkish industrialist, Asil Nadir. A year later the company was seized by British receivers when Nadir, unable to pay debts of $1.6 billion, fled to Cyprus to avoid accounting and theft charges in London. In 1992 the British receivers sold Del Monte to companies run by a young Mexican banker, Carlos Cabal Peniche. When Cabal's empire collapsed in 1994, he too became a fugitive, and the fruit company was taken over by the Mexican government.

In late 1996 Mexico sold eighty percent of its interest in Fresh Del Monte Produce to the IAT Group, a holding company registered in the Cayman Islands and owned by the Palestinian family of Abu Ghazaleh. IAT controls companies which produce and distribute fruit and vegetables in several countries, particularly Chile. In 1997 Fresh Del Monte Produce went public and its shares are now traded on the New York Stock Exchange. In 1998 the company bought the operating subsidiaries of IAT, a purchase which essentially meant buying its parent company. Fresh Del Monte is a transnational company for sure, but for the record, its home base is Coral Gables, Florida.

It's a banana eat banana world out there. All of the banana players that compete in the North American market are multinational corporations with worldwide production and distribution of multiple products. The three U. S. based companies, Dole, Chiquita, and Fresh Del Monte, are the North American and European market leaders and also control about sixty-five percent of the world banana trade. Noboa (Ecuador), whose brand name is Bonita, and Fyffes (Ireland) have about 15 percent of the world market.

Over the past several decades Dole and Chiquita have been in a neck and neck race for the biggest American market share. Since 1996 Dole has had the largest share of the U. S. banana market, but Chiquita remains the world's leading banana producer. Both companies produce a wide variety of food products. With 2000 revenues of $4.5 billion and operations in ninety countries, Dole is the world's largest grower, shipper and marketer of fresh fruits and vegetables and fresh-cut flowers. The Westlake Village, California headquartered company sells almost every imaginable fruit and vegetable including fresh cut salads and pineapple tidbits for pizza. You want fresh satsumas, pomegranates or lychees? Dole has them. Until 2001 the company even controlled 75 percent of the soft drink market and represented Coca Cola in Honduras. Chiquita had 2001 sales of $2.2 billion and distributes fresh and processed products in over sixty countries. The Cincinnati headquartered company is the U. S. leader of private label canned vegetables, a position strengthened by the January 1998 acquisition of Stokely USA, Inc.[386]

Third position in both the U. S. and world banana markets is held by Fresh Del Monte Produce, Inc. It is the largest marketer of fresh pineapples in the world and is the largest marketer of branded off-season cantaloupes in the United States. In 1998 it added grapes, apples, pears, kiwi and other fruits to its produce line. There might be Del Monte fresh tomatoes or carrots in your future.[387]

The fourth largest importer of bananas and the largest importer of plantains in the United States is the Turbana Corporation. Turbana is the U. S. subsidiary of Unibán, a co-op of Colombian growers. The name Turbana comes from a combination of Turbo, a Colombian port city, and banana.

Bonita Bananas from Ecuador are sold in the United States by Pacific Fruit, Inc. Bonita also sells pineapples and plantains from Ecuador and mangoes from Ecuador and Peru. With numerous competitors, the market battle is fought on many fronts.

The scene of competitive battle is the produce department of your local supermarket. Produce is the glamour department of the supermarket. You can see, feel and smell most of the food. Only in produce can you have the tactile thrill of caressing cucumbers, pinching peaches, massaging melons and pawing plums. The produce department is strategically placed based on a supermarket layout more than fifty years old. Since people tend to circle the periphery of a store in a counter-clockwise direction, the high-profit items are placed on the outer walls with main items such as milk and bread in the back left corner. You have to pass everything else to get there. The produce department is usually one of the first areas encountered.[388]

As the major contributor to produce sales and store profit, bananas usually are prominently displayed at the front or rear (you have to pass everything else to get there). Up to ninety percent of produce sales are based on consumer impulse,[389] and an attractive attention-getting display stops at least sixty-six percent of all shoppers.[390] Size also matters. The International Banana Association found that where store displays were increased between 17 and 31 square feet, banana sales by pound increased fifty-eight percent, dollar sales jumped sixty-six percent, and total produce department sales increased five percent.[391]

The banana companies recommend to retailers that bananas be displayed in a single layer on a stepped, padded display. Artificial grass with stiff fibers are to be avoided. The display should be kept fully stocked to avoid a "picked-over" appearance, and overripe or damaged fruit should be removed immediately. Eighty-four percent of customers prefer to buy bananas when they are color #4 to 6 (more yellow than green to all yellow but may have light green necks but no green tips) on the seven step scale. Ray Sonderfan, a discerning New Jersey consumer, says he likes his bananas very mature but firm, whatever color that is. Despite all this attention, bananas are the most profitable produce item partly because they deliver the highest unit sales per hour among all major fruits at the lowest cost per hour to the store.[392]

Something of an supermarket banana display rivalry broke out in recent years. Citing a previous record display of 360 cases of bananas, Vons Pavilion supermarket in Arcadia, California claimed a new record of 421 cases, about twenty thousand pounds, in a four day promotion in 1996. The seventy thousand Chiquita bananas were on sale at five pounds for a dollar.[393] The Vons record was shattered less than a year later in Catonsville, Maryland at a Metro Food Market. The display of thirty-one thousand pounds of Banacol bananas was mounted to celebrate five years of maintaining a consistent price of twenty-nine cents a pound for bananas.[394]

In 1999 a Auchon Hypermarket unit in Houston created a display of twelve hundred boxes. That record quickly fell when Bag-N-Save Foods in Dover, Ohio created a monster display of 1548 cases of Dole bananas weighing thirty-one tons. The display included a life-size bamboo hut on stilts and a trio of mechanical apes playing musical instruments. The monkey band played and sang nonstop while customers picked up bananas on sale for ten cents a pound.[395] Surely, somewhere today someone is planning the mother of all banana displays.

If you have ever had the supermarket checkout clerk mistake your nectarines for peaches, you will understand the value of price look-up codes, or PLUs. Since you can't print the universal product bar-code on bananas, Chiquita in 1994 introduced the now common practice of printing the PLU for bananas—4011—on their little blue Chiquita stickers. It added efficiency and pricing accuracy even if there was no way to confuse a banana with a carrot.

Banana sticker possibilities may be endless. In 1997 Dole decided to make their bananas into billboards when they applied stickers hawking milk and the Twentieth Century Fox movie *Anastasia*. Dole distributed 100 million bunches of bananas with "Got Milk?" stickers in a campaign paid for by a dairy farmer-funded program. For Dole the milk, cereal and bananas breakfast connection was a natural. While Russian Czar Nicholas II and his family were not known banana fanciers, the connection was that the six *Anastasia* character stickers from Dole bananas and other products were redeemable for free movie tickets and toys.[396]

During most of the 1990s the Dole Food Company banana advertising was directed primarily to the youth market. Beginning in 1994 Dole, which markets over sixty fresh fruits and vegetables, used a multi-media approach to promote the "Five-A-Day" servings of fruit and vegetables recommendation of the U. S. Department of Agriculture. A CD-ROM directed at third graders was made available to elementary schools. It used characters such as

Bobby Banana, Amber Orange, Ray Raisin and Bobby Broccoli to get the healthy eating message across. Kids could use an internet site to access nutrition and recipe information and e-mail the fruit and vegetable characters.

Dole's Bobby Banana character went on to become the star of the company's radio and television advertising. The animated TV ads had Bobby, sometimes on a skateboard, singing the Dole B-A-N-A-N-A rock and roll jingle in an animated setting of high kicking bananas and pineapples forming geometric patterns that resembled a Busby Berkeley musical film of the 1940s. Bobby Banana made personal appearances, and for those who couldn't get enough of the banana heartthrob, a Bobby Banana doll was available.[397]

In 2001 Dole launched the Dole Banana Shuffle with a national cable TV campaign. The ad shows Bobby Banana and a group of kids dancing and singing a jingle with the message, "shuffle your way to a healthy day." Available for purchase was a plush toy Bobby Banana and CD-ROMs with nutritional information and how to do the shuffle.

While Dole may focus on the youth market, Chiquita's advertising prime target group is women age twenty-five to forty-five who want to feel better, have more energy, experience a longer life and be good moms. Chiquita's heritage of real people, warmth, humor and fun is also evident in the company's ads. "Message is the most compelling reason why people buy bananas," according to Chiquita's marketing veep, Sherrie Terry, and Chiquita's ads reflect the consumer shift to a healthier life-style.[398]

Chiquita's advertising pitch has traditionally promoted the health benefits of the fruit. In a 1945 movie ad, *Chiquita Banana's Beauty Treatment,* an animated Chiquita banana counsels an exhausted housewife that a daily dose of bananas will lead to a more beautiful appearance and complexion.[399] When the W. B. Doner agency updated the health pitch with real people in 1993, they scouted health clubs and recruited eighty year-old Bert Morrow who bench pressed, did sit-ups and ate a banana every day. Scouting office building lobbies produced Julie Fernando whose daily regimen included climbing sixteen flights of stairs to her office and eating a banana.[400]

Chiquita's 1996 "Real People" campaign featured three athletic types, but an animated Miss Chiquita on the blue label opens and closes each ad. Her blue label transforms into a bicycle wheel in an ad featuring twenty-six year-old Christian Huff who picks up and delivers nearly fifty packages a day, and yes, each morning he opens a little yellow package labeled Chiquita. In another spot Miss Chiquita takes a dip with Jean Durston, an eighty-two year-old, who swims and eats a banana every day. Youth is served with nine year-old Darin Granfield who as a hockey goalie makes a lot of stops including one to eat the banana mom gives him every day. "Chiquita: Quite Possibly the World's Perfect Food" helps Darin get ready for the big boys . . . and the big girls too.[401]

Consistent with its nutritional advertising positioning, in 1997 Chiquita became the first banana company to have its bananas certified by the American Heart Association which entitled the company to use the "Heart Check" mark on packaging, in advertising and on point-of-sale materials. The heart association program requires that a food have less than three grams of fat and no more than 20 mg. of cholesterol per serving. The banana easily qualified.

Where you find bananas, the monkeys can't be far behind. Pacific Fruit, Inc. put the two together in a 1995 regional television and billboard ad campaign to build brand-name recognition for its Ecuadorian bananas marketed under the Bonita label. The TV ads were used in Phoenix, Arizona and Portland, Oregon and featured singing and dancing monkeys that urged shoppers to "Eat a Bonita." "Morphing" techniques made the live monkeys appear to be talking and singing.[402] Apparently, women aged twenty-five to forty-nine, the target group, were thought to be quite taken with monkeys.

Sometimes well intended banana promotions flop. The 1991 "Chiquita Value Plan" was such a case. Customers could call an 800 number, and after providing a nutritionist with demographic information about their family and what they spent on food, they received a personalized menu plan. Actually, there were only two possible menu plans, the economic and—for higher incomes—the liberal. Not surprisingly, it was suggested that a weekly buying plan include seven pounds of bananas. An outside nutritionist found fault with too much cheese and meat in the weekly plan, and because there were no per ounce serving suggestions, thought there was too much potential for a fatty diet. A privacy watch-dog thought that soliciting family demographic information was "insidious."[403]

How do you fit a nine inch banana in a kid's eight inch lunch box? You don't; you get a shorter banana. That was partly the idea that launched the "Chiquita Jr." in 1992. Chiquita realized the regular nine-incher was often too much for young children and older people. Retailers who test marketed the juniors in Cincinnati and New Orleans reported that the smaller bananas were well received but probably cannibalized sales of the full sized fruit.[404] Today the "Chiquita Jr." has found a selective group of customers and a bipolar market—kids and seniors—around the country. The junior bananas come packed in their own pink and blue boxes and have a PLU sticker different from their bigger siblings.[405] Even junior sized innovations can help in a competitive business.

As global concerns about environmental issues increased in the 1990s, all of the banana companies have been under pressure to address such concerns in both the way that they do business and in the image they project to the customer. Banana production raises many environmental issues because of the high use of pesticides, disposal methods for plastic and organic waste, chemical contamination of water and exposure of farm workers to chemicals.

Europe has been known for "green" movements for decades, and it is there that a movement that respects labor and environmental standards caught on. It is known as "Fair Trade" and its origins lie the 1960s and 1970s when "trade, not aid" was seen as a development tool. In relation to bananas "Fair Trade" is seen as an instrument to improve conditions for banana farmers and plantation workers and to protect the environment. In 1988 Solidaridad, a Dutch development organization, worked with the Max Havelaar Foundation in Switzerland to introduce fairly traded coffee to the consumer market. Cocoa, tea and honey followed. In November 1996 fair trade bananas arrived in Europe with an introduction in the Dutch market. They came from an environmentally managed plantation in Ghana and a small farmers' organization in Ecuador. Within one month the "Ok*" branded bananas gained a ten percent share of the Dutch market. In March 1997 the fair traded bananas arrived in the Swiss market and captured a thirteen percent market share within three months. As one publication put it, the "Ok*" brand "put the cat among the pigeons of the two big US-based multinationals, Chiquita and Dole."[406]

The small Ecuadorian producers who supplied the fair trade bananas to Switzerland received a minimum guaranteed price of $7.75 per box which covered all of the costs of production and processing and included a premium of $1.75 per box for investments in social infrastructure, ecological improvements, business development and human resource development. A survey in the European Union by the European Commissioner of Agriculture showed that seventy-four percent of the population would buy fair trade bananas if they were available in stores alongside regular bananas and thirty-seven percent would be willing to pay at least ten percent more for them.[407] These developments forced the big multinationals to give increased attention to their human resource and environmental practices as they sought to expand existing markets and develop new ones.

In an industry noted for some unethical if not illegal business activities in the past, Chiquita won considerable praise when it released its *2000 Corporate Responsibility Report.* The Report candidly admitted Chiquita's shortcomings in the past while identifying progress made in correcting environmental and social problems in the areas where it does business.[408]

In 1992 Chiquita made a commitment to a farm certification program created by the Rainforest Alliance. Originally the program was known as ECO-O.K., but it is now known as the Better Banana Project. The Rainforest Alliance in partnership with the Conservation Agricultural Network, a network of environmental groups in Latin America, aims to bring operating procedures in developing countries up to par with those in wealthier countries.

Companies that wish to participate can achieve crop certification by meeting environmental and labor standards. According to Mauricio Ferro, head of the Corporation for Conservation and Development, "At the most basic level, we're trying to teach producers that the river is not a dump and that employees are human."[409] Spot checks of farms monitor compliance.

Each year beginning in 1994 Chiquita increased the number of its owned farms certified to the standards of the Better Banana Project. By 2000 Chiquita reported that one hundred percent of its 127 company owned farms employing about 20,500 workers and covering 65,500 acres in Colombia, Costa Rica, Guatemala, Honduras and Panama had achieved Rainforest Alliance certification. Some thirty percent of the independent farms that provide bananas sold under the Chiquita name also achieved certification.

Chiquita reported that over eight years it spend some $20 million dollars to meet criteria requiring that farms responsibly manage agrichemicals, conserve wildlife habitat and natural resources, and promote worker welfare and community well-being. For instance, the Better Banana Project standards prohibit the use of the "Dirty Dozen" pesticides which is a group that consists of particularly harmful and persistent organic pollutants. Chiquita reports that it uses only pesticides that have been approved for use on bananas by the United States Environmental Protection Agency and the European Union regulatory authorities.[410]

The Dole company also adheres to environmental standards in its operations. In 1998 Dole in Costa Rica became the first banana exporter and first agricultural producer in the world to have its environmental management systems certified under ISO 14001, a set of requirements established by the International Organization for Standards. All of Dole's farms that grow bananas in Latin America and Asia are now certified to the ISO standard. While dealing with similar environmental concerns, ISO 14001 is different from the Rainforest Alliance standards. Dole says it uses the same "best management practices" internationally as used in its United States operations.[411]

Beginning in 2001 Dole began selling organic bananas in the United States. The bananas are grown in Ecuador and Honduras on farms that have been certified as organic by independent US-based auditors and inspected by the Independent Organic Inspectors Association. Although organically grown bananas can be twice as expensive as the traditional banana in the supermarket, that is not always the case.[412]

It appears that all the banana companies recognize that good environmental practice is good business. Fresh Del Monte's environmental policy explicitly states "that sound environmental management is critical to our business success."[413]

As political change swept Eastern Europe in the 1990s, the American banana companies saw fertile new ground, not to grow bananas but to sell them. While the full potential has not been reached, there have been definite changes. "PIG Export—Import," a Polish banana marketer notes on its web site that fifteen years ago finding bananas in Polish shops was a miracle. Now the exotic fruit is easily bought.

The potential of the Russian market was demonstrated in the early years of the Yeltsin administration when Chiquita offered to send ten million bananas as a Christmas holiday gesture of good will to the hard pressed residents of Moscow. The original idea was to give bananas to the entire population of the Russian capital, but when local authorities indicated that a food riot might result, the bananas were directed to poor people and school children. Chiquita spokesperson Magnes Welsh observed the distribution and noted that the children, many of whom had probably never seen the yellow fruit, had various ways of eating the bananas, but they were all "gone in a nanosecond."[414]

Other marketing horizons beckon as well. In the last quarter of 1997 Chiquita began marketing bananas in China for the first time. Both Dole and Del Monte also now have a presence in that country. Imagine the potential of more than one-fifth of the Earth's population awaiting the golden fruit.

CHAPTER 8

The Great Banana War

"The future isn't bleak.
There's always going to be people eating bananas."
Cameron Forsythe, Puerto Armuelles Fruit Co., Panama,
2001

A "Great Banana War" developed in the 1990s that spanned three continents and involved dozens of countries but, fortunately, was fought with political, legal and economic weapons. It involved fallen colonial empires, and, unlike past world wars, the United States and Germany were on the same side.

The conflict broke out in 1993 when the European Union (EU), dedicated to free trade amongst its member states, imposed stricter quotas and tariffs against banana imports from Latin America except those from former EU colonies known as the African, Caribbean, and Pacific (ACP) states. The EU policy aimed at cutting Latin American banana imports to the EU from 2.7 million to about 1.35 million tons each year. While Europeans would get fewer bananas at a higher price, the policy was intended to assist the ACP countries such as Jamaica and Cameroon, among others.[415]

The new EU banana import regime was a significant change because in 1992 before the new restrictions sixty-two percent of the EU's bananas came from Latin America. The overseas territories of EU members supplied twenty percent and the ACP states only eighteen percent of the EU's banana needs. The big three

United States companies alone supplied about 43.5 percent of EU consumption in 1991.[416]

Among the EU countries, Germany was the most outraged over the new banana restrictions both because Germany had no former colonies to assist and because the Germans were positively obsessive about bananas. "In Germany, bananas are an impossibly overdetermined symbol, signifying justice, national self-determination, cultural pride, deprivation, prosperity, Communist tyranny, capitalist luxury, unity, and economic and even sexual freedom."[417] No wonder the Germans went bananas over the issue. The German obsession goes back to the early years of the twentieth century. Bananas, introduced to Germany in the decade prior to the First World War, became plentiful and cheap, but the supply was cut off for five years during the war. In 1919 a correspondent of the *London Times* found a crowd of Berliners gazing with awe at a bunch of bananas hanging in a delicatessen shop. "The smiling faces and little jokes made it quite evident that the banana was recognized as a symbol of peace and that the delight felt at its presence was due to the evidence it afforded that the [wartime] blockade is a thing of the past."[418]

World War Two once again cut off the banana supply but not the German taste for the fruit. This point was made somewhat strangely in the 1981 German film *Das Boot (The Boat).* The film, based on an autobiographical novel by Lothar-Guenther Buchheim, was written and directed by Wolfgang Petersen. It is one of the most revered antiwar films in film history and follows the crew of forty-three men in U-96 in their battles against the British navy. We see in the film that after a considerable time at sea, the sub docks at the neutral port of Vigo, Spain to refuel and reload torpedoes from an interned German supply ship. Mountains of green bananas are brought on board and piled so high that they must be moved before navigation charts can be rolled out. The boat is ordered to enter the Mediterranean through the seven mile wide Strait of Gibraltar under the noses of the British navy patrols swarming through the area. As we watch the surfaced boat glide silently into the strait at night, one officer is seen calmly munching on a banana. The mass of bananas which have ripened considerably since last seen are now hanging conspicuously like so many Christmas decorations throughout the control room. When the boat is spotted and all hell breaks loose, the bananas are in the thick of the action as the boat dives two hundred eighty meters to the bottom. Both men and bananas survived the depth-charges.

Unlike the film, a real report from Vigo, Spain on September 11, 1940 raises questions about German loyalty to the banana. On that day Vigo's fishermen put to sea to pick bananas. They discovered that large areas of the water were covered with bananas. A German ship had thrown overboard more than five hundred tons of them. The fishermen sold the fruit at low prices.[419] It is not known if German submariners were among the customers.

As children, post-war generations of Germans, other than mariners, knew about the yellow fruit only through the recollections of their elders; thus the fruit brought memories of deprivation and humiliation. Bananas came to symbolize luxury to both West and East Germans. In the 1950s West Germany became the single largest importer of bananas in Europe. In 1957 West German Chancellor Konrad Adenauer filibustered for four days to insert a guarantee of tariff-free bananas in unlimited quantities in the treaty that established the EU and then praised the fruit as "paradisiacal manna" in a passionate speech to the German Bundestag. In communist East Germany bananas were nonexistent after the war, and their absence was viewed as a symbol of communist failure to provide the simplest of luxuries.

When the Berlin Wall fell in 1989, East Germans saw the banana as a symbol of freedom and German unity. By 1991 the former East Germans were packing away fifty-six pounds per person per year. The West Germans, at thirty-one pounds per capita, were no banana slackers. East Germans became so enamored with bananas that it became part of the national psyche. *Der Stern* reported in a 1990 survey of East German sex shop sales that a banana shaped vibrator and a banana-flavored condom known as the "Wild Banana" led all other brands. Politicians in the East attracted crowds to rallies by handing out free bananas, and their West German compatriots began calling the Easterners "Bananen."[420]

Considering the German fixation with bananas, the EU's imposition of tariffs on Latin American bananas in 1993 inevitably produced a "banana-split" in the EU. On the free-market side was Germany, Belgium, the Netherlands and Luxembourg; the three large producers of Latin American bananas, Chiquita, Dole and Del Monte; and the "banana republics" that produce the large, bright yellow Latin American "Dollar Bananas." Favoring the tariffs and quotas were the ACP states, producers of the smaller and paler "Eurobananas," and their former colonizers such as France and the United Kingdom. They were joined by Spain, Portugal, Italy and Greece. Some observers saw the split going beyond mere bananas. Some parties thought the battle was also to keep Germany from throwing its weight around.

Germany appealed the banana tariffs to the European Court of Justice, but the Court rejected the appeal in 1994. At about the same time the EU ruled that beginning in January 1995 all imported bananas except those from several European former island colonies must be "free of abnormal curvature" and be at least 5.5 inches in length and 1.1 inches in diameter. The Germans were bent out of shape over both decisions. People on both sides of the issue began insulting their rival's bananas with such terms as "skinny," "tasteless," "easily bruised," and"rotten."[421] There were even charges of "banana-laundering" as some importers were accused of trying to disguise the true source of their bananas.

Some German firms began to look for ways around the new regulations. In late 1993 Mercedes-Benz, the German vehicle manufacturer, signed a barter agreement with Ecuador. Mercedes-Benz contracted to exchange one hundred of the company's buses for $6 million worth of bananas in order to circumvent the EU trade restrictions. The company planned to sell the bananas in eastern Europe.

A Hamburg fruit importer, T. Port GmbH & Co., announced in 1995 that it planned to build greenhouses to grow bananas on the German Baltic Sea island of Rugen, an island known for its bone-chilling winters. Some German newspapers thought the idea was preposterous. A spokesman for the company said, "This is a banana war. We are fighting an unjust and inefficient European policy. In some cases , the tariffs are three times as high as the worth of the bananas." Experts said the high cost of heating the green houses in winter made the idea impractical. The Rugen islanders welcomed the idea of new jobs but thought the fruit importers had gone bananas.[422]

A spokesman for a German fruit importing firm noted that, "The Americans are our natural allies in this war. After all, if the EU gets away with this policy on bananas, Brussels [EU headquarters] might try the same thing with pineapples or peanuts later."[423] How right he was, and there was no bigger ally than Chiquita Brands. From 1991, pre-restrictions, to 1994, post-restrictions, Chiquita's European market share dropped from 25 percent to 18.5 percent, and according to a Chiquita spokesman, when the surplus fruit was dumped on the United States market, Chiquita lost more than $400 million from lower banana prices.[424] Not surprisingly, Chiquita went to the United States government for relief by filing a complaint with Mickey Kantor, the U. S. trade representative in the Clinton administration, and by launching a lobbying campaign in Congress.

In 1996 President Clinton attended a black-tie benefit for Washington's historic Ford's Theater where he was entertained by a high-class vaudeville style show including luminaries such as Natalie Cole, Gregory Hines, Jon Bon Jovi and Bill Maher. Among the performers was popular comic Paula Poundstone who displayed a journalist's inquisitiveness about the evening's seating arrangements. From the stage Poundstone greeted the President and Mrs. Clinton seated in the front row. Questioning the woman responsible for the seating arrangements, Poundstone wondered how seating decisions had been made? "Money" was the response. And who was the gentleman seated directly behind the President? Getting no response, Poundstone asked the gentleman directly. Still no response. Then the president stated awkwardly, "It's a secret!" The audacious Poundstone continued her probing asking the gentleman how he got so wealthy. The woman in charge of seating then informed the audience the gentleman in question was the "Chiquita Banana Man."[425]

Carl Lindner, a.k.a. the "Chiquita Banana Man," had made large campaign contributions to former President George Bush and Republican causes such as GOPAC. Now he was hobnobbing with a Democratic President. *The Wall Street Journal* observed in 1995 that it was unlikely that Mr. Lindner had seen the ideological light. Instead it seemed that "He's seen who turns on the light in the Oval Office and is trying to buy influence there."[426]

The influence may have bought the Clinton administration's opposition to the EU's banana import policy. The United States would normally have little interest in the issue since only a handful of jobs were at stake and the United States doesn't export bananas. Between 1993 and 1996 Lindner and Chiquita donated more than $1.1 million to the Democratic Party. The contributions were legal, and both sides denied that there was any quid pro quo. Besides the choice seating at Ford's Theater, it is known that Lindner was also one of the more than nine hundred who got coffee in the Oval Office and a sleep-over in the Lincoln Bedroom. "Lindner wanted the administration to do a lot more than we did," said Kantor in April 1997. Lindner wanted the U. S. to impose sanctions on the EU; Washington filed a complaint with the World Trade Organization on April 11, 1996.[427] The action was somewhat rare. Hundreds of companies ask Washington to investigate unfair trade practices, but the U. S. Trade Representative accepts only about fourteen cases each year. Only a few of those are accepted by the World Trade Organization for resolution. When Kantor was asked at a news conference to identify any American economic interests at stake in the banana war, he thought awhile and, smiling broadly, replied, "If you drive north on the Pacific Coast Highway, I think there's ten acres of bananas on the right-side."[428]

The effort to aid Chiquita was truly a bipartisan affair. Perhaps no one was more supportive of Chiquita's case than Republican Senator Bob Dole. Senator Dole was not directly connected to bananas despite the fact that Dole Foods Co. sells bananas. This lack of a connection apparently wasn't well known in Turkey where the Mayor of Izmir in western Turkey announced in February 1995 that because Senator Dole was waging an intensive campaign against Turkey and was the owner of Dole bananas it would not be in the national interest to buy Dole bananas. Izmir's municipally owned supermarkets said they would replace Dole bananas with those of Chiquita and Bonita. Two days later the Mayor of Izmir admitted that he had been misinformed about Mr. Dole, but he claimed that his action demonstrated that his country could not be pushed around by the United States. The sale of Dole bananas resumed immediately.[429]

While Senator Dole may not have owned Dole bananas, it seemed in 1995 and 1996 that Chiquita owned Senator Dole. In the Fall of 1995 the Senator unsuccessfully tried several times to include amendments in the budget bill that would impose sharp trade sanctions on Costa Rica and Colombia because of deals they had cut involving their banana exports to the European Union. The two countries had worked out a special arrangement for licenses to export to the EU under the new quota restrictions. Chiquita contended that these deals further eroded its European market. To bolster his position, Dole got a pledge of support from House Speaker Newt Gingrich and said he was acting because the Clinton administration had failed for two years to resolve the case of trade discrimination. Since Dole's home state of Kansas grows a lot of corn but no bananas, observers were surprised by Dole's persistence on the banana issue. Washington insiders contended that the cause of Doles' ardor was contributions from the Lindner family and corporate interests to the Republican Party and the Dole presidential campaign. The Senator also benefited from frequent use of Lindner corporate aircraft while campaigning.

When Senator Dole's efforts failed to bear fruit, Lindner took a direct hand in trying to change Colombia's policy by meeting with that country's President, Ernesto Samper Pizano, in a Miami hotel room in December, 1994. The banana man showed pictures of himself with Ronald Reagan and George Bush and bragged of his relationships with Gingrich and Dole. Samper was both shocked and angered by the incident. Despite the pressure from Lindner and Capitol Hill, Mickey Kantor ruled out sanctions against Costa

Rica and Colombia even though both countries were found to be in technical violation of U. S. trade laws.[430]

Colombia was not entirely off the hook. In yet another connection with the Great Banana War, the Clinton administration in early 1996 decided to remove Colombia from the list of countries cooperating satisfactorily with international narcotics control efforts, an action known as decertification. There were legitimate reasons to do so. Colombian President Samper was under a cloud of suspicion due to allegations that he received millions from Cali drug traffickers for his 1994 campaign. Some U. S. officials were claiming that Samper also obstructed drug-control efforts by Colombia's national police. Some media observers believed that Clinton's punishing of Colombia through decertification was more of an effort to curry favor with big contributor Lindner as the 1996 presidential elections approached.[431]

What is known is that Lindner dropped a bundle in the 1996 election season. The day after Trade Representative Mickey Kantor filed a complaint with the WTO, the banana king and his top executives began contributing more than $500,000 to Democratic party interests in more than two dozen states. This was a procedure recommended by the Democratic National Party for "hot potatoes," defined as donors who wanted some anonymity. There was nothing illegal about this procedure, but was it a money-for-favors deal? *New York Times* columnist William Safire thought so and chose to label the affair "Bananagate." A Chiquita spokesman said, "The charge that Carl's [Lindner] contributions were linked to policy changes simply does not reflect reality." Kantor said the case was based on recommendations by U. S. Trade Representative civil servants, not Lindner's campaign contributions.[432]

Ecuador, Mexico, Guatemala and Honduras joined the U. S. complaint to the WTO, but the action was bad news for many Caribbean nations such as Jamaica, Suriname and the Windward Islands (St. Lucia, Dominica, St. Vincent and the Grenadines, and Grenada) which benefited from the EU preferences. These Caribbean nations found some support in Washington including an unlikely source, a four-star U. S. general. Marine General John

Sheehan, commander of all U. S. forces in the Atlantic and the Caribbean pointed out that the eastern Caribbean drug trade could explode if the banana growers on the Windward Islands lose the European market. Speaking at a forum in Washington, the general said he was "not taking a position on international trade." The general noted that the islands are dependent on a single crop: bananas. If the islands were to lose their EU preferences for their higher cost bananas, Sheehan said that people would resort to drug dealing and illegal migration to provide for their families. Representative Maxine Waters (D-Calif.), another forum participant, agreed and said, "These countries will be destroyed by the drug lords. They lose their only export and it's over." Former Dominica Prime Minister Dame Eugenia Charles echoed the same opinion. "If we lose our banana industry, it will be very difficult to tell young people not to go into drugs."[433]

On March 18, 1997 the WTO in a preliminary ruling found that the EU's banana policy violated global trade rules. "We won the banana case? Well, I guess the meaning is that it helps to be Carl Lindner," said Gary Hufbauer, a trade economist at the Institute for International Economics.[434] For Alfred Prosper the reaction was a little more personal. Mr. Prosper, 53, has been growing bananas for twenty years on a six acre plot that he owns on the island of St. Lucia. "Why is America doing this to us? This is a little place, and all that we know, and what we depend on. We have nothing else and we hurt nobody, but now they want to take even this from us." Prosper is part of the one-quarter of the labor force that is employed in some way in the banana industry on the island nation.[435] When a U. S. trade delegation visited St. Lucia in April of 1997 they received an even more forceful attack in the banana war. Meeting with the managing director of the St. Lucia Banana Growers Association, Rupert Gajadhar, the delegates listened to a thirty-minute tirade in which Gajadhar accused the Clinton Administration of a "diabolical" campaign saying, "You are conducting the worst kind of economic warfare against a defenseless people. You take away our bananas and leave us with no alternative but misery, strike, suffering." It wasn't quite the pleasant Caribbean escape from the Washington political wars that the Americans expected.[436]

The official WTO ruling in favor of the American position in the Great Banana War was issued in May, 1997, just in time for a visit to the Caribbean by President Clinton. Meeting in Barbados with the leaders of fifteen Caribbean countries the banana issue took center stage. Prime Minister P. J. Patterson of Jamaica noted "That for many of our countries, bananas are to us what cars are to Detroit." The meeting did not offer any more hope for Caribbean bananas in Europe than American automobiles might expect in Japan. The meeting did produce "Partnership for Prosperity and Security," a document characterized by one observer as "flowing diplobabble." The statement promised further discussions and recognized the "critical importance" to the Caribbean nations of continued access of their bananas to the EU.[437]

The EU appealed the WTO ruling and lost. Winners such as Ecuador noted that Latin American farmers using more mechanization could produce bananas at around $160 per ton while the Caribbean's smaller farms with more human labor produced bananas at about $500 per ton. The Germans were looking forward to their big "Dollar bananas" while Britain promised to find some way to fight the WTO ruling. Meanwhile in the Caribbean, St. Lucia was looking to attract a replacement crop of tourists while in Dominica and St. Vincent, both with more rain and fewer beaches, farmers were reported to be diversifying into marijuana.[438] Without European preferences the prospects for Caribbean banana growers look bleak. Without major gains in production efficiency or tapping new markets—air-freighted bananas to China was considered—the Caribbean nations can hardly be expected to compete with the industry giants.[439]

And what of Chiquita? In a press release of September 9, 1997, the company expressed extreme gratification with the WTO decision and looked forward to recovering the fifty percent share of the EU market it had lost. But Carl Lindner was taking no chances. Between January 1997 and August 1998 Lindner and his wife, Edyth made contributions of $360,000 to Republicans and $176,000 to Democrats, sums which ranked Lindner number four in the "MOJO 400" rankings of political contributors.[440] After Lindner met with Speaker Gingrich in October 1998, Gingrich and Senate Majority leader Trent Lott met with the U.S. Trade Chief Charlene Barshefsky to prod the White House to push for faster compliance with the WTO ruling.

The 1997 WTO decision did not end the war over bananas, that most political of fruits. In January 1998 a WTO arbitrator gave the EU until January 1, 1999 to comply with its ruling. The EU proposed modifications which the complainants, the U.S., Ecuador, Guatemala, Honduras, Mexico and Panama, found unacceptable. In June of 1998 the European Agricultural Council of the EU adopted modifications to the EU banana measures and unilaterally declared them to be WTO consistent. The EU rebuffed U.S. led efforts to negotiate and prompted the U.S. to develop retaliatory measures in late 1998. A full-scale trade war threatened.

The Economist pronounced the retaliatory move as absurd monkey business. The magazine wondered if the WTO's dispute settlement mechanism was ineffective and noted that both the United States and the EU were slow to implement WTO rulings. The EU was dragging its feet in allowing imports of hormone-treated beef from America and the U.S. was slow to lift its ban on importing shrimp from countries using nets that trap turtles.[441] *The Economist's* concern was prompted by the release in November 1998 of the list of potential European exports that would be subject to punitive one hundred percent U.S. tariffs. The initial list included such things as German coffee-makers, toy trains, windshield wipers, communion wafers, Christmas baubles and "practical joke articles" which one pundit speculated might include exploding bananas.

By December 1998 the fury surrounding the banana issue became almost surreal. While the United States House of Representatives debated impeachment of the president and bombing strikes were taking place in Iraq, President Clinton was engrossed in discussions over bananas with leaders of the European Commission.

With Sir Leon Brittan, the trade minister of the European

Union, holding forth on the issue the President finally remarked that the only thing that made the discussion tolerable was listening to the Brit pronouncing "bo-non-nas."[442]

Failing to win EU concessions, the United States announced in January 1999 the list of European imports on which it would impose $520 million in punitive import duties on February 1. There were forty-two types of products on the tariff increase "hit-list." U.S.-based companies that would be affected immediately launched a campaign of public-hearing testimony and lobbying. Neiman Marcus bemoaned the impact on the price of Italian and Scottish cashmere sweaters, and Louis Vuitton hoped that its handbag market would survive the price increase. Sweet biscuits, pecorino cheese and prosciutto, chandeliers, greeting cards, bath products and bed linens also made the hit list. A Scottish newspaper speculated that the United States made up the list by sending someone on a walk down Madison Avenue to write down the names of items in boutique shop windows.

When the final list of goods destined for the high tariff was published in April 1999, only nine types of products were covered. Most of the goods on the original list had been stricken. American merchants selling such items as industrial batteries, English decorative lithographs and German bath products saw the cost of those goods double. Such businessmen without political clout were destined to be, as the military might put it, collateral damage in the Great Banana War.[443]

The EU protested the U. S. action and tried to stall a WTO meeting to consider the U.S. position. The EU requested arbitration over the $520 million amount of the duties, an amount that the U.S. claimed represented the loss to American companies arising from the EU's banana regime. The final tariff list imposed duties on $191 million of European imports, an amount the WTO found to be justified.

After protracted negotiations, the United States and the EU reached an agreement in April 2001 to end the nine year Great Banana War. The tariff and quota regime set up in 1993 is supposed to be replaced by 2006 with a tariff only system. During a transition period companies will receive licenses to fill import quotas based on their import volumes to the EU from 1994–1996. The last point left Dole—the banana company, not the Senator—bent out of shape. Dole would have preferred a "first-come, first-served" transition arrangement. While Chiquita steadily lost market share in the EU, Dole was diversifying its banana sourcing to adapt to the now overturned regime. Under the new plan, Dole's efforts counted for little while Chiquita, with the largest volumes between 1994 and 1996, should benefit the most from the transition arrangement.[444] The EU formally adopted the changes in December 2001.

There were casualties in the Great Banana War. There is an old African proverb that says, "When the elephants fight, the grass gets trampled." Grass was trampled in this conflict; but then again, much of it was the same grass that has always been trampled in the banana business, banana workers. When the EU imposed its new banana regime in 1993 and the US banana companies lost a large part of their most-profitable market, the companies were forced to adopt survival strategies such as reducing production and lowering costs. Banana workers were in the front lines of the battle strategy.

In September 2001 more than three thousand workers in Panama were on strike to protest cost-cutting measures by Chiquita subsidiary, Puerto Armuelles Fruit Co. Banana picker Roberto Chacon Sanchez voiced the complaint of the workers. "More work, less money. I have been doing this all my life—I will die here—and all my life the big banana companies have been saying they are not making money and will leave. They haven't left yet. If you believe them, in twenty years they haven't made a dime." Company cost-cutting included closing three farms and combining several jobs into one. To Chacon "cost-cutting" also meant "wage-slashing" because adding more responsibilities slowed down individual worker production, and employees were usually paid by volume.[445]

The Puerto Armuelles strike was settled in October 2001 after four months of negotiations. Workers accepted changes in production levels and working conditions without a decrease in their wages. The company reported that the company's personnel had been reduced by seven hundred workers during the year.

Part of the problem for many Latin American countries and their workers is that business was moving to the lowest cost producers. With the lower prices and oversupply of bananas created by the EU restrictions, cheaper producers such as Ecuador are driving high-cost producers such as Costa Rica out of the market. Unlike most Latin American producers, in Costa Rica fifty-eight percent of the production is from independent and mostly small-scale farms. In 2002 Costa Rican banana workers were earning about $18 a day which was almost double the minimum wage. They also had social security benefits and subsidized electricity, health benefits and housing. Although Costa Rican productivity was the world's highest, its price of $5.20 a box of bananas was just break even. At the same time Ecuador sold its fruit for $2.18 a box and paid workers less than $6.00 a day with few extra benefits. Luis Umana of the Association of Independent Banana Producers in Costa Rica noted that, "Costa Rican bananas are not like others; they come with many social guarantees. We need to make this known to consumers."[446] Nevertheless, it is no surprise that the big-three US companies all canceled some contracts with Costa Rica's independent producers at the time.

Ecuador produces about thirty-six percent of the world's exported bananas and supplies close to one-quarter of all bananas consumed in Europe and the United States. But Ecuador's success may come at a high human price. In an April 2002 report issued by Human Rights Watch, the organization took Ecuador to task in reporting that children as young as eight work on the banana plantations and that adult workers are not permitted to organize unions. The organization interviewed forty-five children and found that forty-one of them began work on plantations between the ages of eight and thirteen. The average work day was twelve hours and fewer than forty percent continued attending school past age fourteen.[447]

A *New York Times* story in July 2002 also reported a significant child labor problem on banana plantations in Ecuador. Esteban Menendez is a ten-year-old who works with his father who tends ninety-eight acres on a three thousand acre plantation owned by Alvaro Noboa, Ecuador's richest man and a candidate for the country's presidency in 2002. Esteban climbs banana plants and ties insecticide-laced cords between them to prevent them from collapsing under the weight of the produce. His situation is fairly typical in a country where families have to send young children to work to survive. The International Labor Organization estimated in 1999 that sixty-nine thousand children ages ten to fourteen were working in Ecuador, and the situation may be even worse in other countries in the region.[448]

Ecuador supplies about twenty-five percent of the bananas eaten in the United States. The bananas from Mr. Noboa's plantations are marketed in the United States mostly under the Bonita label. Ecuador is an increasingly important source of bananas for Del Monte and Dole particularly since the Great Banana War caused companies to look for ways to reduce costs. Chiquita does not own any farms in Ecuador and purchased only a small part—five percent—of its bananas there in 2000.

Dole states that it "does not knowingly purchase products from any commercial producers employing minors. It is Dole's policy to observe local labor laws."[449] Chiquita addressed the child-labor issue candidly in their *Corporate Responsibility Report* by stating that their Code of Conduct prohibits the use of child labor. The report also noted that it was found that one contractor in Guatemala had employed several fourteen-year-olds for temporary project work on company-owned housing, a practice which is permitted under Guatemalan law. Chiquita also stated that "because employees generally live in communities near the farms, children are often present in and near work areas, although they themselves are not working. In recent years, the Company has taken action to ensure that children do not accompany their parents into the fields or packing stations." In any case the Company planned to have all of its divisions adopt formal policies describing how they would respond if children were found to be working for the Company.

Such policies are required by the SA8000 standards that Chiquita follows.[450]

The Ecuadorian government claims it wants to fight chronic child-labor abuses on its banana plantations, but the Labor Minister Martin Alberto Insua says the problem is that Ecuadorian producers are paid "unfair" prices. A forty-three pound box which the Ecuadorian producer may sell for less than $3 may bring $25 or more at retail in the United States or Europe with the largest portion of that going to the retailer. One cause of this situation is the consolidation of some huge retail outlets such as Wal-Mart and Costco which can almost name the price they will pay the distributor. Thus, everyone all the way back to the field worker feels the pinch since it is a highly labor-intensive business with a product that in March of 2002 sold in the United States for about fifty-two cents a pound.[451]

While the Great Banana War did not cause questionable environmental or labor practices by the banana companies, the companies were forced to address those issues in a business environment made more difficult by the banana fight with the EU.

One of the combatants in the Great Banana War was dealt an almost mortal blow by the fruit battle. Chiquita Brands International, Inc. suffered losses in seven of nine years before 2001. With a stock price of $18 a share in 1997, Chiquita shares dropped to a low of forty-two cents in 2001. In January 2001 the company was unable to pay interest on $862 million in debt and accrued interest and began to negotiate with its bondholders to restructure its debt. Chiquita in November 2001 entered into a prearranged Chapter 11 bankruptcy plan of reorganization. In March 2002 the bankruptcy court approved a plan by which creditors owed nearly $700 million received ninety-five percent of the company's stock. Carl Lindner's American Financial Group which had owned thirty-one percent of the common stock of Chiquita on December 31, 2001 owned less that one percent after the reorganization. Although Lindner retained a seat on the board, he had stepped down as CEO in 2001.

Although the banana war with Europe was a major contributor to Chiquita's problems, some claim that poor business decisions played a part. In the early 1990s Chiquita put all of its bananas in one basket, namely the belief that the European Union would liberalize its banana trade policies. Instead, the EU tightened the trade with its quota and tariff system. Chiquita says it lost half of its European market overnight.

With Carl Lindner at the helm in the late 1980s, Chiquita configured itself to exploit the European market. Between 1987 and 1991 sales in Europe grew at eight percent per year and constituted more than half of the company's banana sales. Believing that the EU would remove trade restrictions and expecting to penetrate the newly liberalized economies of Eastern Europe and the former Soviet Union, the company planned for expansion.

Between 1989 and 1993 Chiquita increased its production capacity by thirty-two thousand acres in Latin America and took delivery of fourteen refrigerated ships. Most of this was done by adding debt which by 1993 reached 270 percent of equity, a highly leveraged position. When the Europeans imposed their new banana regime, the company discovered "it had a surplus of bananas and nowhere to sell them; a fleet of ships and nowhere to sail them; and a mountain of debt and no way to pay it."[452]

A former Chiquita executive recently remarked, "If a secretary on the 27th floor of the Chiquita building got pregnant out of wedlock, the company would blame it on the European Union.[453] Since the company attributed it poor performance to the trade policies of the EU, Chiquita decided to challenge the new quota system to regain its European market share. Meanwhile both Dole and Del Monte, also affected by the EU policy, decided to make the best of it and adapt to the new circumstances.

Both Dole and Del Monte were also bruised by the EU banana regime, but both chose to diversify and expand into new markets. They also bought into banana operations in the Caribbean and Africa that gave them preferential access to European markets.

Dole bought a sixty percent interest in a Swedish vegetable, fruit and flower importer, and also acquired a Spanish produce grower. The company also bought some European importers that held licenses to import bananas under the new regime. One European diplomat observed that "Dole pursued a smart business strategy, while Chiquita pursued a vendetta."

Del Monte, already a leader in fresh pineapples, expanded its melon and other fruit operations. Mohammad Abu-Ghazaleh, Fresh Del Monte's CEO, commented that, "I don't want to hang my mistakes and my bad management on somebody else. We will just have to adapt to the reality. We are in business, not politics."[454]

Chiquita is now under new management. Dole and Fresh Del Monte are fairly well diversified. People will always eat bananas so the business prospects seem unlimited. But bananas have always been political, and in an age of multinational business, banana politics is both domestic and international. That is not likely to change. Fortunately, countries have not yet taken to building arsenals of ICBMs, intercontinental banana missiles.

CHAPTER 9

The Healthy Banana

Little Miss Muffet
Sat on a tuffet
Eating her curds and whey;
But, surely, you know,
That was long years ago—
She feeds on Bananas to-day.[455]

Thursday, September 5, 1996 was a day that American tennis star Pete Sampras will remember in his U. S. Open experiences. It wasn't so much his quarterfinal victory over Alex Corretja; it was the manner in which the fifth set tie—breaker was accomplished. Sampras found himself retching in a far corner of the Louis Armstrong Stadium Court before walking away a winner in a four hour, nine minute record setting ordeal that required two IVs to restore his body's energy. Afterward, Sampras commented. "During the third and fourth sets, I was starting to drink a little bit of Pepsi, which wasn't the smartest thing to do. I felt I needed a kind of caffeine, but it was just dehydration. Maybe I should have had a banana . . . I just ran out of gas." Banana or no banana, Sampras survived to go on to defeat Goran Ivanisevic in the semis and bested Michael Chang in the final to claim his fourth career Open title and his eighth career Grand Slam singles crown.[456]

Dr. Hans Jakob Rogler, a thirty-five year old vegetarian, in 1930 walked and ran 375 miles from Oslo, Norway to Stockholm, Sweden in ten days subsisting solely on a diet of bananas and milk. He ate six ripe bananas and about a quart of milk each day and finished in excellent physical and mental health.[457]

Today each runner in the New York City Marathon receives a banana when they finish the race. Runners are partial to bananas although sometimes there can be too much of a good thing. Race organizers of a 5-K race in Seattle in 1999 had twenty-one hundred runners registered, but discovered on race day that enough bananas had been delivered to feed twenty-one thousand runners. The runners were invited to help themselves to the virtual mountain of bananas.[458]

The Dallas Cowboys trainer once reported that the pro football players at the training table were partial to broccoli, cauliflower, oranges and bananas.[459] Cyclists are known to prefer bananas as an en route snack. Is there a pattern emerging here? Yes, bananas are good for athletic performance, but you don't have to be a runner or football player to enjoy their nutritional benefits.

Bananas are possibly the world's most perfect food, as Chiquita likes to point out in its advertising. An average 3.5 ounce (100 gram) peeled banana provides 22 grams of complex carbohydrates and natural sugars that your body needs for energy. About twenty-two percent of the banana is carbohydrate ranking it second in carbohydrate content among forty-two common fruits and vegetables. Only the sweet potato has more carbohydrate content.[460] Carbohydrates are stored as glycogen in your body, and glycogen is broken down as needed into glucose, the fuel the body needs to run. A glycemic index ranks foods on a scale of 0-100 based on how quickly they are converted to usable energy. With a glycemic index of 55 the banana gives a quicker energy boost than other common fruits. The orange has a rating of 44 and an apple a rating of 38.[461]

When the San Francisco 49ers opened their NFL season against the Buccaneers at Tampa, trainer Lindsy McLean prehydrated the team for days in preparation for Florida's temperatures in the 90s and humidity near seventy percent. Diuretics such as Coke, coffee and alcoholic drinks were out; fruit juice, Gatorade and water were the drinks of choice. Lots of fruit was available, and McLean particularly pushed bananas.[462] Why? Potassium! Bananas are high in potassium, a key mineral lost during physical activity. One banana contains about 450 milligrams of potassium, about eleven percent of what is needed each day and the best among fruits. You would need to consume about three apples to equal the same amount of potassium. A potato with its skin has about 850 mg. of potassium, but they are not nearly as convenient as the banana. Who would want to eat a potato *au naturel'* as a mid-afternoon snack? Potassium controls muscle contractions, helps keep a healthy heartbeat, and control's the body's water balance.

Bananas provide about twenty percent of the U. S. Recommended Daily Allowance (USDRA) of Vitamin B6 and fifteen percent of Vitamin C, and they are a good source of fiber. They have about one milligram of sodium, the lowest among fruits, and contain virtually no fat or cholesterol. At about one hundred calories per medium banana, no wonder the banana "smiles." So for all those who suffer from orthorexia, the abnormal fixation of eating healthy foods, the banana is right up your alley.

But bananas do have their nutritional detractors. A study of the nutritional value of common fruits by Paul Lachance, Ph.D., of Rutgers University found that the kiwi ranked highest among fruits in the ability to provide the recommended amounts of nine essential nutrients. Using an index which calculated the daily value of the nutrients per one hundred grams of the fruit, the kiwi was highest at sixteen. The banana had an index of four, ahead of the apple at two.[463] Nevertheless, the kiwi with its dark green fuzzy skin and an interior of green fruit and black seeds won't win a beauty contest with the banana.

Long before Americans discovered the kiwi they were singing the praises of the banana. Sample the testimonial of Mr. John C. Sleater of Merchantville, N. J. written to the Fruit Dispatch Company in 1919. "My breakfast consists of a shredded wheat biscuit moistened with hot water to which has been added a tablespoonful of peanut butter, one slice of crisp toast, two bananas and the juice of an orange or grapefruit in hot water. I can work for five or six hours after this meal without further desire for food." Perhaps such a menu would not be everyone's choice of breakfast fare today, but as he said, "I know from practical demonstration that the banana is one of the best vitalized foods on the market."[464]

The nutritional quality of the banana can make it a handy fruit in a crisis. In December 1996 rebels who were part of the Tupac Amaru Revolutionary Movement in Peru seized the Japanese ambassador's home during a gala dinner. Diplomats and businessmen were taken hostage. Two days later, with 375 male hostages placing a heavy demand on the food, clothing and sanitary facilities of the ambassador's home, the Red Cross was called upon for aid. While TV cameras recorded the scene, aid workers delivered 800 bananas, 450 apples, mineral water, toothpaste and toothbrushes, razors, medicine and clothing. Note that the bananas were delivered before the portable toilets; someone had their priorities straight.

In 1952 the editorial page of the *New York Times* extolled the virtues of the banana while noting that there was something strangely affable about the fruit. The editors observed how easy it was to peel a banana and concluded that comparing the banana with "the resistant orange, the difficult lobster and the fortified coconut, it may well be that the banana does find some sensuous satisfaction in being ingested by man." With tongue firmly planted in cheek, the editors itemized a litany of abuses inflicted upon the uncomplaining banana. People fry them, drown them in milk, and spoil their natural flavor with cane sugar. Mothers mash them for babies and children roughly skin them. The newspaper concluded that placing the manifest affability of the banana against these abuses, "it might be well to consider whether it is not time to acknowledge that the banana and not the dog is man's best friend."[465]

Bananas have been recognized as man's friend for centuries, particularly in China. A Chinese writer in A.D. 713 recognized that bananas eaten raw had the medical effect of quenching thirst, "lubricating" the lungs, stopping bleeding and healing wounds. In powdered form bananas were thought to stimulate circulation of the blood, strengthen the marrow in bones and cure fevers. In the twelfth century A.D. Fan Ch'êng-ta wrote that bananas were good for feeding babies and could be preserved by soaking in sugared plum juice, drying and pressing flat. In his book on herbals, published in 1590, Li Shih-chên enumerated 1892 drugs and cited preserved bananas as a delicacy in the North. The Chinese regarded banana root as a cure for abscesses, fever, jaundice, measles, and headache with fever. Banana oil was also regarded as somewhat a miracle drug. As a hair tonic it was thought to stop women's hair from falling out and to help hair grow long and dark. It was also seen as a cure for epileptic fits if the patient could be made to drink it and vomit.[466] What more can be asked of such a humble fruit?

Perhaps inspired by the recent American interest in alternative medicine, a delegation of psychiatrists and pharmacologists recently journeyed to India to examine what is known as Ayurvedic medicine, a form of humoral medicine. Humoral thinking goes back at least to Hippocrates (460-327 B.C.), the father of medicine, who wrote "Let thy food be thy medicine and thy medicine be thy food." The basic doctrine of Ayurvedic medicine is that there are four basic humors of the body; blood, black bile, yellow bile and phlegm. There are four qualities of sensory experience: wet, dry, hot and cold; and four ingredients of things going in and out of the body: air, fire, water and earth. To be healthy all of these factors must be in harmony, and some fine-tuning is necessary. Hot must be balanced with cold, wet with dry and so on. A humoral physician's job is to assist the natural culinary processes with warmers and coolers and to facilitate evacuation processes with emetics, bleedings and purges. You are what you eat and what you excrete.[467] Thus it is no surprise that the banana could have medicinal uses in India. The plantain, the banana's close cousin, has a huge folklore reputation as a treatment for peptic and duodenal ulcers. The Indian *Materia Medica* (1954) indicates that green plantain flour made into a handmade bread is good in cases of dyspepsia and flatulence, and a slight gruel of banana flour with milk is easily digested in cases of gastritis.[468]

Equatorial African traditional healers use bananas as a herbal remedy. For fast energy bananas are carried on the hunt and on rigorous journeys. The seeds of some varieties are ground to a fine powder and taken as a laxative. When mashed, including the skin, and mixed with boiled milk, they are a native remedy for diarrhea, colitis and ulcers. A salve of mashed bananas combined with honey is used as a treatment for hemorrhoids, a case where the sticky cure may be worse than the condition. Inner banana skins are applied as poultices for healing burns, wounds and boils, and in areas where "Band-Aids" may be few and far between, bandages are often made from dry stem portions of the banana plant. With all these banana remedies and use of the banana as the main starch in the diet, the banana provides a lot of "bang for the buck," . . . or naira . . . or ekuele.[469]

Henry M. Stanley wrote of his 1874 explorations in Africa in *Through the Dark Continent.* His party of sixty-two men at times found it hard to obtain food, but on several occasions bananas obtained locally saved the day. Of one experience he said, "bananas, ducks and coffee, and the tobacco gourd and pipe close one of the most delicious evenings I have ever passed." He also observed that "[bananas] when ripe form an admirable dessert, and taken in the morning before coffee, serve with some constitutions as an agreeable laxative." Overall, Stanley marveled at the usefulness of the banana. "The banana plant will supply a peasant of Uganda with bread, potatoes, dessert, wine, beer, medicine, house and fence, bed, cloth, cooking pot, tablecloth, parcel-wrapper, thread, cord, rope, sponge, bath, shield, sun hat, even a canoe, in fact, almost everything but meat and iron. With the banana plant he is happy, fat and thriving; without it, he is a famished, discontented, woe-be-gone wretch, hourly expecting death."[470] Not much more could be expected from the humble banana.

Americans, too, find some ingenuous uses for the banana. Bananas as a cure for headaches has received attention in Ann Landers' columns. One Greensboro, North Carolina woman reported how she licked a killer migraine. "I used commercial duct tape to keep half the banana peel on my forehead and the other half on the back of my neck. In less than 30 minutes my headache was gone."

Her chemistry teacher husband attributed her cure to a potassium deficiency that the banana filled. A Chatham, Ontario reader told Ann Landers that she went into banana cure mode when she felt a headache coming on. When she saw herself in the mirror with banana peel taped to her head, she could not stop laughing for ten minutes. She then took a nap and woke up without the headache. "I don't know if it was the banana peel, the laughter, the nap or the combination of all three, but from now on, there will always be a banana in my refrigerator."[471] Did you hear that Miss Chiquita, "in the refrigerator?" In the interest of full disclosure, bananas have also been known to cause migraines in some people.

Some American banana cures are a little more grounded in medical practice. In 1929 the *Boston Globe* reported the wondrous cure of little Barbara Ann Chisholm who had been wasting away in her crib unable to retain anything on her stomach. Local physicians and the best specialists were perplexed. All known remedies had utterly failed. The child was removed to the Children's Hospital in Boston where for months a search was made for some diet that would restore the child's health. The food was finally found—ripe bananas!

A small portion of mashed ripe banana was tried and agreed with the child. Eventually a diet was worked out that eliminated all sugar, starch and fats. Banana remained the principal food. After six months in the hospital the child was taken home. After more than a year on the banana diet, a healthy, happy, smiling Barbara Ann had more than doubled her weight. Doctors expected that by the time she was five years old, she would be able to eat normally.[472] The ailment was diagnosed as celiac disease, also known as sprue.

Celiac disease is an autoimmune intestinal disorder disturbance due to the inability to utilize fats and carbohydrates, especially a protein called gluten found in many grains, in a normal manner. The use of carbohydrates in the form of ripe bananas was reported in medical literature as early as 1924.[473] Some people felt that doctors often misdiagnosed the disease, but when the banana diet was introduced, it was looked upon with wonder by some doctors.[474] Some people may have carried the cure to extremes. It was reported in 1954 that a young boy in Seattle, who was stricken with celiac disease when he was eighteen months old, had been cured by eating a diet of lean meat and bananas. He ate seventeen thousand bananas, approximately three tons. It was not reported who kept score, but it is known that the young man still enjoyed bananas after he was reportedly cured.[475]

When World War II made bananas almost a museum rarity, especially in northern cities, the parents of babies suffering from celiac disease often had to resort to public pleas for bananas. In mid-1942 bananas were rushed by plane from New York to Montreal for a twenty-two month old baby. The mother of another young sufferer published an open plea in the *New York Journal American*. Brooklyn police searched for hours to find twenty-four bananas and raced them to the aid of a fifteen month old baby. The United Fruit Company announced it would give priority to celiac sufferers.[476]

A New Englander named Bill remembers the experience of his older sister. Joyce was born in 1939 and was a "normal" child, but it was noted that she was slow to gain weight and failed to "thrive." In April 1940 at the age of thirteen months the "old" family physician diagnosed celiac disease. Joyce was put on a diet of bananas, cottage cheese and canned milk, a regime she followed until 1946.

When the war broke out and bananas became scarce, the doctor petitioned the federal government to provide a supply of the fruit. Bill recalls, "Out city was just outside of Boston. We would take the train into town and then walk to the piers on Atlantic Avenue. The sight of my mother followed by two small children often evoked smiles and laughter for the men there. But they understood the importance that bananas played in Joyce's well-being. They saved her life." There was a benefit to Bill too. If there were extra bananas and it was just before a shipment, "I was allowed a rare, delicious treat—a mashed banana with sugar and milk," says Bill.

Sometimes the bananas were flown in, and Bill tells the story. "One of my earliest memories is that of going to a place where the bananas were being flown to. I must have been only four. Travel by car was unusual as gas was rationed. Travel by car at dusk was even more unusual. We arrived at a grass field and waited. When the sound of the plane's engines was heard, my dad and one other driver there present turned on the headlights. We had a 1936 Hudson and the lights were prominently mounted on the fenders. They were large, and the brightness surprised me. I was not used to seeing them, and just a few months before I had made a terrible mistake. Our house was located directly on the water front. The air raid sirens had sounded; the fear was that enemy submarines were in the area of the channel between Boston Harbor and Fore River shipyard. I lifted the shade at the window to see the submarines. Perhaps my first memory was that of the old English-born warden storming up onto our front porch and then reading me the riot act. He had survived the bombings of London, and I heard what terrible fate could and would befall us all if I ever raised the shade or 'leaked light' again."

"Anyway, Bill recalls, "the plane landed and my dad and this other man removed the crates from the small plane, placed them in the back seat with me, and we drove away as the plane prepared to take off. I was very glad to leave that place. Years later, I remember having seen the response from the Federal Government detailing the procedures that would be followed in providing bananas for my sister. Even then, it caused a slight racing of my heart."

On January 18, 1947 John McCann, Lord Mayor of Dublin, Ireland, where the tropical fruit was a rarity since the early days of World War II, sent a cablegram to Mayor O'Dwyer of New York which read: "Poet Eoghan Roe Ward dying. Bananas may save life. Is it possible send some by air?" Two hands of bananas were sped to the airport and placed aboard a Pan-American flight for Shannon. When the bananas reached the bedside of the Irish poet and ballad singer, doctors said he was too ill to eat them. The Irish bard died on January 23 from his digestive ailment without ever eating the bananas. Oh, yes, Ward was his pen name; his real name was Patrick Michael O'Connell.[477] Fifty pounds of bananas flown to a two year old boy in Lockerbie, Scotland in 1950 produced better results.[478]

Continuing with gastroenterological concerns, the BRATT diet has long been a remedy for intestinal upset and diarrhea. Consisting of bananas, rice, applesauce, tea and toast, the diet has virtually no fat that will create turbulence in the GI tract. The components of the diet are easy to have around; may be eaten in a variety of combinations; and are more nourishing than various tablets and pink potions from the medicine cabinet.[479] "So if your stomach is bothering you, a banana a day may keep the doctor away!"[480] A word of caution however. Unripe bananas cannot be digested in the small intestine because they contain resistant starch. The starch thus ferments in the large intestine causing what might be a bloomer buster wind. Remember: "They must be ripe!"

Bananas have long been used in folk medicine to build strong stomachs. Indian medical studies using unripe plantain powder induced healing of duodenal ulcers in seventy percent of the patients. A placebo had the same effect in sixteen percent of the cases. The banana stimulates the proliferation of cells and mucus that form a stronger barrier between the stomach lining and corrosive acid. British researchers found that "the role of bananas in folk medicine as an antiulcerogenic agent, at least against gastric ulcers, appears justified." An Australian animal study using very ripe bananas apparently produced a substance that armor plated the stomach's protective lining and reduced the body's secretion of stomach acid. Bananas may possibly one day replace expensive drugs in ulcer treatment.[481]

Bananas can be helpful in controlling hypertension. A study at the Institute of Internal Medicine and Metabolic Diseases of the University of Naples in Italy found that eighty-one percent of the subjects with high blood pressure who daily ate three to six servings of potassium-rich foods such as bananas at the end of one year could cut their hypertension medication in half. A joint University of California at San Diego and University of Cambridge School of Medicine study showed that people who ate a single serving of a potassium rich food such as a banana had a forty percent lower risk of stroke.[482] A study published by the American Heart Association in 1998 supported such findings. An eight year study of forty-four thousand men found that those whose diets included large amounts of potassium had one-third fewer strokes than those who did not.[483]

Scientists at Johns Hopkins University and the National Heart, Lung, and Blood Institute pooled the results of more than thirty well conducted trials on potassium and blood pressure published since the 1920s. They concluded that people who have high blood pressure can lower it by increasing their consumption of potassium; high potassium consumption may prevent the development of hypertension in the first place. Some of the studies utilized a dietary approach to lower blood pressure. The added benefit of eating more fruits and vegetables rich in potassium, including bananas, is that they are also low in sodium, virtually fat-free and high in fiber. Most Americans average about 2,500 milligrams of potassium daily. In the studies analyzed by the Johns Hopkins researchers, the addition of about another 2,340 milligrams kept blood pressure down. That's the equivalent of about five bananas.[484] Of course no one should undertake a high potassium regimen or a reduction of blood pressure control medicine without consulting their doctor.

No discussion of medical studies is complete without knowing what happened to the rats. Indian investigators found fiber from unripe plantains caused the blood cholesterol (LDL)—the bad kind—to plunge to one third of previous levels and raised HDL cholesterol—the good stuff— by about thirty percent. It is likely that ordinary bananas also lower blood cholesterol because of high pectin content which in bananas is higher on a weight basis than in apples, a confirmed cholesterol lowering food.[485] Lest you think we tout the banana too much, there are other foods high in potassium. You might also try avocados, lentils, molasses, nuts, parsnips, canned sardines and fresh spinach among other choices. In any case, if bananas were your only source of calories, a person would have to eat twenty-four a day.[486]

Bananas rarely cause allergic reactions. A four-year old British boy is probably grateful for that fact. After hundreds of tests the boy was found to be allergic to ninety-five percent of foods. Unfortunately, the youngster also suffers from asthma, eczema and epilepsy. Dr. Alan Smyth of Nottingham said that the young man may have to spend all of his life on a diet of broccoli, cauliflower, chips, bread and bananas because doctors can not find a cure for his allergies.[487]

Bananas, or at least their smell, may eventually play a role in weight control. Scientists at the Smell and Taste Treatment and Research Foundation, Ltd. discovered that the brain can be fooled into thinking you are full by inhaling various scents when hungry. Director Alan R. Hirsch, M. D. said "The odors act on the satiety center in the hypothamus, or the area of the brain that makes you feel full." Some three thousand overweight subjects used inhalers to take three sniffs of food fragrances of either banana, peppermint or green apple whenever they felt the urge to snack. Sniff frequency ranged from 18 to 285 sniffs per day. Subjects lost an average of five pounds per month.[488] Before you start sniffing bananas you should know that the *Tufts University Health and Nutrition Letter* questioned the validity of Dr. Hirsch's claims and pointed out that his claims have never been subjected to the scrutiny of fellow scientists.[489] Meanwhile sniffing your bananas can't hurt.

Then there are those who prefer to smoke banana peels. In the counter culture of the 1960s some among the flower generation smoked banana peels in a variety of forms in place of or in addition to marijuana. Many credit the Donovan Leitch song "Mellow Yellow" with encouraging the banana smoking practice. Donovan claims that the words "electric banana" in the song was in reference to new sex toys that were coming on the market at the time. San Francisco folk-rocker Country Joe McDonald said that just a week before the release of "Mellow Yellow," it was he who started the scam about getting high by smoking banana peels. Apparently, the process produced little effect beyond suggestion. Elaborate recipes for preparing banana peels for smoking by microwaving, scrapping, smashing, mashing, drying and mixing can be found today on the Internet, and the "cooks" swear by the effects although one does caution that chromosome damage may lead to men growing breasts. However, the total result may be less than the sum of the efforts. Bananas, particularly the peels, do contain minute amounts of psychoactive substances such as serotonin, norepinephrine and dopamine.[490] These chemicals are also produced naturally by the body and are believed to relieve mental depression, but smoking banana peels is not a recommended healthy procedure.

For some survivors of the infamous World War II "Bataan Death March" held by their Japanese captors at the Cabanatuan Camp in the Philippines, smoking banana peels was a last resort. Robert Body, a private from Detroit, remembers the prisoners addicted to tobacco. "They'd smoke anything they could get. The Red Cross cigarettes were the best. But it got to where some of these guys would smoke weeds, leaves, even banana peels."[491]

Perhaps the banana's most significant positive contribution to the health of the world's population lies in the future in the form of a vaccine delivery system.

More than three million children in developing countries die each year from diseases such as hepatitis B, cholera and diarrhea. Most children in countries such as the United States are vaccinated against such diseases at a cost of $50 to $100 per child. Such costs are prohibitive in most of the developing world. What if kids could be vaccinated simply by eating a banana?. Plant molecular biologist Dr Charles Arntzen and researchers at the Boyce Thompson Institute for Plant Research in Ithaca, New York are seeking the answer to that question. In 1992 the World Health Organization (WHO) launched the Children's Vaccine Initiative as a cooperative effort to promote vaccine research for the developing world. During a trip to Thailand that year, Dr. Arntzen noticed a mother feeding her infant a taste of banana. At that moment it became obvious to him; the banana would be the perfect food for an oral vaccine. Bananas are cheap, widely available and often fed to children. What could be more simple and painless than eating a banana? And very significantly, bananas are often eaten uncooked; the vaccine would not be destroyed by heating. The question then became how to get the vaccine into the banana.

The biotechnology research technique being used is the insertion of genetic material from a virus or bacteria into the DNA of plants. Then the genetically altered plants are grown. Theoretically, a person would be vaccinated by eating the plant material. With traditional vaccines, chemically damaged or weakened viruses or bacteria are injected into people. The body's immune system responds, but because the invaders are weakened they don't cause disease. The body then builds up an immunity so that the next time the real disease enters the body, people don't get sick. With an oral vaccine the injection would be eliminated.

Researchers at Roswell Park Cancer Institute in Buffalo, New York and at Texas A & M University tested the theory by inserting DNA from hepatitis B virus into tobacco plants. Mice inoculated with leaf extracts from the plants showed the full immune response seen in human vaccinated in the usual way. These researchers and Arntzen's group at Boyce experimented with potatoes engineered with hepatitis B, *Escherichia coli* (e. coli), and Norwalk virus. The latter two are responsible for numerous cases of diarrheal disease in developing countries. Mice showed some immunity after eating the potatoes.[492] Now the goal is to get vaccines into bananas that would be grown locally in developing countries. Arntzen estimates that approximately twenty thousand tons of bananas would be needed to vaccinate all of the world's children. This number represents a mere fraction of the nine to ten million tons produced in the world each year. One banana could contain as many as ten doses of vaccine. Another advantage of bananas is that they can be eaten uncooked thus preventing vaccine denaturation. Medicinal bananas might even have a different colored peel if the pigmentation gene is altered. Imagine that; yellow for normal bananas, purple or magenta for the medicinal ones.

The Boyce Institute in 1997 after success with mice tested human volunteers with raw potato vaccines against cholera and diarrhea caused by e. coli bacteria. In 1998 Dr. Arntzen announced that this first trial on humans was a success. The volunteers who ate small samples of genetically engineered potatoes developed an immune response to traveler's diarrhea, sometimes known as "Montezuma's revenge."

Clinical testing began in 1999 on a vaccine genetically engineered into a potato to provide immunity against the Norwalk virus which annually afflicts an estimated twenty-three million people in the United States alone. Nineteen of twenty volunteers who ate the transgenic potatoes developed an immune response to the virus. Animal and human studies so far have provided proof of the idea that edible vaccines are feasible, but there are still many issues to be addressed.[493]

Getting the correct amount of the right protein to grow in a food is an issue. Proper dosing could be another problem. What is the right amount of the fruit vaccine? It seems unlikely that it can be as simple as having a village shaman distribute bananas. Probably the vaccine will have to be processed into something like a freeze-dried banana-chip and then distributed by medical professionals.[494]

Funding the research and development of edible vaccines is a problem which consumes much of Dr. Arntzen's time. The big pharmaceutical companies aren't beating down his door with fists of money. They get a very tiny percentage of their profits, perhaps one percent, from vaccines. Since the first successful splicing of plant DNA in 1983 more than forty different species of genetically engineered food have been produced. But there is considerable outright resistance to the idea of genetically modified crops, and it may be that the technical obstacles may be more surmountable than the political ones. While edible vaccines, bananas or otherwise, have a long way to go, it would certainly be gratifying to have a relatively inexpensive system to improve the health of millions of people worldwide.

While people may some day eat bananas for medical reasons, today they are consumed because they are good for you, and people like them. Some ways in which bananas are eaten have become almost legendary either because of the popularity of the preparation or the status of the banana devotee'.

His Royal Highness Edward, Prince of Wales revealed his favorite breakfast while visiting the better watering places of the French Riviera and Southern Spain in the 1930s. "At breakfast time peel a banana. Lay it on a plate of ample size and ladle over it thick orange marmalade. Eat with a spoon, munching at the same time hot buttered toast. Sip coffee."[495] His Royal Highness certainly had plenty of time to munch and sip.

Perhaps the most remarkable banana breakfast of all time is a fictional one. It occurs in the novel *Gravity's Rainbow* by Thomas Pynchon. CNN has described Pynchon as "an enigma shrouded in a mystery veiled in anonymity." To say he shuns publicity is an understatement. Until a *New York Magazine* reporter tracked him down in 1996, no reporter had interviewed him in more than four decades. There are no known photographs of him since the early 1950s.[496] His picture was also missing from his freshman year Cornell directory when he arrived there as an engineering major in the early 1950s. Some of his fans have wondered if Pynchon is a pseudonym for a committee, a computer, Woody Allen or even Professor Irwin Corey, the disheveled, zany, unintelligible TV comic.[497] CNN concluded that he actually leads a rather conventional life in New York City, so conventional that you might not know him even if you saw him.

Gravity's Rainbow, which has been granted equal stature with *Moby Dick* and *Ulysses*, was the co-winner of the National Book Award for 1973. The three members of the Pulitzer Prize jury unanimously recommended the fiction award for Pynchon's *Gravity's Rainbow*. However, other members of the fourteen-member board turned down the recommendation and no fiction prize was given in 1974. Board members described the novel during their private debate as "turgid," "unreadable," "overwritten," and in parts "obscene." One editor said he tried hard but had only gotten a third of the way through the 760 page book.[498]

Length may be the least problem of what one critic called "the most widely celebrated unread novel of the past thirty years." When the editors of the *New York Times Book Review* selected *Gravity's Rainbow* as one of the three outstanding books published in 1973, they said it was "one of the longest, darkest, most difficult and ambitious novels in years . . . bonecrushingly dense, compulsively elaborate, silly, obscene. funny, tragic, poetic, dull, inspired, horrific, cold and blasted."

Bananas dominate the opening of *Gravity's Rainbow* set during World War II. Captain Geoffrey "Pirate" Prentice, who has the weird ability to manage the fantasies of other people, is famous for his banana breakfasts and messmates from all over England throng to his London maisonette near Chelsea embankment. Pirate collects the ripest bananas "among these yellow chandeliers" from his rooftop greenhouse while watching a German V-2 rocket, "a steel banana," make its way toward London.[499] Pirate, driven to despair by the wartime banana shortage, got a friend who was flying a route to South America to pinch a sapling banana tree or two in exchange for a German camera.

Pirate prepares his banana feast while his drinking companions slowly assemble; Teddy Bloat staggers into the kitchen and slips on a banana peel, a typical Pynchon use of slapstick. The mob gathers round a great table crowded with

> banana omelets, banana sandwiches, banana casseroles, mashed bananas molded in the shape of a British lion rampant, blended with eggs into batter for French toast, squeezed out a pastry nozzle across the quivering creamy reaches of a banana blancmange to spell out the words *C'est magnifique, mais ce n'est la guerre* (attributed to a French observer during the Charge of the Light Brigade) which Pirate has appropriated as his motto . . . tall cruets of pale banana syrup to pour oozing over banana waffles, a giant glazed crock where diced bananas have been fermenting since the summer with wild honey and muscat raisins, up out of which, this winter morning, one now dips foam mugs full of banana mead . . . banana croissants and banana kreplach, and banana oatmeal and banana jam and banana bread, and bananas flamed in ancient brandy Pirate brought back last year from a cellar in the Pyrennes also containing a clandestine radio transmitter . . . [500]

Bananas are everywhere in *Gravity's Rainbow*. One critic found that "Chiquita Banana's phallus arches over *Gravity's Rainbow* from the beginning of the novel..to the end."[501] It is near the end that Tyrone Slothrop, the book's main character, finds himself in a refrigerator fantasy

> Feelin' yellow and bright as you skirt the bananas, gazing down at verdigris reaches of mold across the crusted terrain of an old, no longer identifiable casserole—*bananas!* who-who's been putting bananas—
>
> In-the-re-frig er a-tor!
>
> O no-no-no, no-no-no!
>
> Chiquita Banana sez we shouldn't! Somethin' awful'll happen! Who would do that? It couldn't be Mom, and Hogan's in *love* with Chiquita Banana . . . . [502]

A considerably less obtuse menu with bananas appears in Iris Murdock's *The Sea, The Sea* wherein a retired theatrical director, living alone in the country, describes a lunch consisting of anchovy paste on hot toast, baked bean and kidney beans with chopped celery, tomatoes, lemon juice and a "really good" olive oil. " . . . Then bananas and cream with white sugar. Bananas should be cut, never mashed, and the cream should be thin." That is followed by biscuits and New Zealand butter, cheese and most of a bottle of Muscadet. "I ate and drank slowly as one should (cook fast, eat slowly) and without distraction such as (thank heavens) conversation or reading. Indeed eating is so pleasant that one should even try to suppress thought."[503] Indeed, the banana is a quiet fruit discretely fulfilling its purpose in life with no hint of a snap, crackle or pop.

Eating bananas in reality can sometimes be stranger than fiction. Such occurred in September 1917 in Bellingham, Washington where one John Frye, a woodsman, ate thirty bananas and then went to a restaurant where he ordered a double portion of halibut, a steak and all the extras. Afterward, he said it was nothing for him to eat a dozen pies at one sitting.[504]

Another American with an even more prodigious appetite was none other than Elvis Presley who died—allegedly died to some—in 1977. Nutritionists have concluded that Presley would be dead by now in any case as a result of an artery-clogging diet. Elvis never met a calorie he didn't like. He was known to down a dozen cheeseburgers at a time and top them off with a gallon of ice cream. The King could down four pounds of burnt-to-a-crisp bacon as a snack. In a one night binge he is reported to have consumed thirty cups of yogurt, eight honeydew melons and $100 worth of ice cream bars. Among the King's all time favorites were fried peanut butter and banana sandwiches. Two tablespoons of peanut butter and one ripe mashed banana were spread between slices of bread. Margarine was heated in a frying pan over medium heat. Browned on both sides in the melted margarine, the sandwich was eaten with a knife and fork while still warm.[505] Registered dietitian Chris Rosenbloom estimated that such a sandwich has 815 calories and 60 grams of fat, almost an adult male's recommended 65 grams of fat per day.[506] One of Elvis' fabled eating exploits included eating nothing but peanut butter and banana sandwiches for seven weeks.[507] Should Elvis stop in at the "Elvis Presley Memphis Restaurant" which opened on the twentieth anniversary of his "alleged" death in 1997, he will find his favorite sandwich on the menu.

In honor of Elvis, all Presleyterians keep thirty-one items in the house just in case of the King's second coming. Known as the thirty-one commandments, these are items which were available at all times at Graceland and in kitchens wherever Elvis went. This list of essentials included twenty-three food items such as: the makings for hot dogs, hamburgers and meatloaf; beverages like orange juice, Pepsi and orange drink; ice cream, brownies and fudge cookies; and Feenamint laxative gum. Prominent on the list was the King's favorite dessert, "`naner pudding." which was to be made fresh each night.[508] When a bloated Elvis was hospitalized in Memphis in 1975 under the care of a dietitian, he persuaded his nurse to make him a batch of banana pudding. Elvis followed breakfast with vast quantities of banana pudding spooned straight from the serving bowl.[509]

While Elvis liked his banana pudding simple and straight, it may be that there are as many ways to prepare the dish as there are bananas. *Many Ways of Cooking Bananas*, a booklet published in

London in the early twentieth century by the Irish firm of Elders & Fyffes, offered no less than thirty-six banana pudding recipes which seems to have confirmed the company's own statement that "the possibilities of the Banana are well nigh infinite."

Elvis Presley and bananas made it onto the University of Chicago admissions application form in 1997. An optional question in the writing portion of the application posed the following scenario. "Elvis is alive! OK, maybe not, but here in the Office of College Admissions we are persuaded that current Elvis sightings in highway rest areas, grocery stores and Laundromats are part of a wider conspiracy" involving among other things the Mall of America, the crash of the Hindenburg, Heisenberg's uncertainty principle, lint, J. D. Salinger and wax fruit. "Help us get to the bottom of this evil plot."

A female applicant from St. Louis, postulated that *Catcher in the Rye* author J. D. Salinger was actually Jesse Garon Presley, Elvis' twin, previously thought to have been stillborn in 1935. She wrote that Jesse was paranoid that Elvis' fame and overall phoniness would distract from his own artistic genius. Jesse changed his name to J. D. Salinger and became a recluse. Today, Elvis appears at bookstores to promote Salinger's novels. He leaves behind wax bananas to remind the world of the King's favorite dish, fried peanut butter and banana sandwiches.

The idea of the Elvis essay was to demonstrate creativity by making quirky and improbable connections, a skill that is thought to be the mark of a University of Chicago student. Presumably, the prospective student made the grade; it is not known if she shares the King's love of peanut butter and bananas.[510]

New Orleans was an early banana port and became a center for famous banana desserts. The *Vieux Carre*, the French Quarter, is the birthplace of some of New Orleans' greatest dishes. "Bananas Foster," a signature dessert at Brennan's Restaurant, is a creation of the late 1940s. Owen Brennan, an Irishman who established the restaurant in 1945, was looking for a way to set his establishment apart from the other French restaurants with a high culinary pedigree. He chose to promote "Breakfast at Brennan's" and, supposedly inspired by the omnipresent bananas seen on the Mississippi River docks, decided that a banana dessert would be a grand finale to "Breakfast at Brennan's." With that inspiration Brennan's talented Dutch chef, Paul Blange, created the dish made with ripe bananas, cinnamon, sugar, butter and brown sugar which is then twice flambed, first with rum and then banana liqueur. The hot banana mixture is served over a scoop of vanilla ice cream. The dessert was named after a regular Brennan's customer, Richard Foster, owner of the Foster Awning Company.[511]

Banana cream pie is another popular American confection and again a New Orleans version stands out. The pie is the signature dessert of Emeril's Restaurant in the "Crescent City." At $7.50 a slice, a quality product is to be expected, and pastry chef Louis Lynch doesn't cut corners. Instead of mixing the bananas into the filling, the layers of sliced bananas are layered with the pastry cream made of egg yolks, sugar, vanilla, heavy cream and cornstarch. The pie is very firm because the pastry cream is made the night before from a whipping cream of forty-percent butter fat. The crust is made with graham crackers, sugar, butter and crushed bananas. The bananas are very visible in the beautiful presentation.[512] Healthy it isn't, but everyone is entitled to go bananas once in a while.

Some banana desserts may be diet busters, but bananas may become the basis for fat-free baked goodies. In the summer of 1997 Chiquita Brands introduced a Fat Replacement System that can replace one hundred percent of the fat and fifty percent of the sugar in baked goods. The fat replacement system is composed of dehydrated banana flakes, cellulose gel and cellulose gum. The banana flavor completely disappears when the material is hydrated and baked. According to a Chiquita spokesman, "If you want a banana flavor, you have to add it." Designed for commercial baking, the product is all natural, has a natural color for light color baked goods and contains high levels of potassium. A baker who tested the product reported excellent results with brownies and cookies.[513] Don't be surprised if your blueberry muffins soon contain bananas.

In other parts of the world bananas can be found in other guises. A Philippines company, Tentay Food Sauces, makes banana ketchup, a concentrated sauce made from fresh, mature bananas with vinegar, spices, salt and sugar. It can be used in other sauce preparations and as a dip for pasta, cold-cuts, hot-dogs, and fried dishes. The Perth Winemaking Club in Australia uses very ripe bananas, including the skins, grape concentrate and other ingredients which, with lots of tender loving care and about seven months, produces banana wine.

A banana by any other name is a plantain, and it is this cousin of our banana that is a staple starch in much of the Southern hemisphere. They have nutritional qualities similar to the banana, but the plantain is a little higher in calories. Compared to bananas, plantains are usually larger and have a more mottled and rougher skin. Plantains blacken when they are fully ripe, but they can be cooked at any degree of ripeness. With a higher starch content then the banana, they are used as a vegetable. Eating plantains raw is not recommended unless they are very ripe. Plantains can be boiled, baked, grilled, sautéed and mashed. As America becomes more multicultural, the plantain is likely to become more than an oddity in North American cuisine.

If bananas were eaten only for the nutritional and hunger satisfying qualities, they might be dull indeed. Some people believe that bananas have magical energies that can improve your spiritual, romantic and even financial health. To those who dabble in such things, the spiritual magic of bananas is associated with the banana's elemental ruler, air. Foods thus ruled are generally useful for strengthening the conscious mind. In addition to the banana, some other foods so endowed include beans, olives, rice and kumquats. If you wish to pursue a spirituality diet, burn white or purple candles and visualize spirit energies while peeling and cutting the banana. Cut the banana into circles, a form which represents the spiritual world. To achieve the maximum benefit, immerse yourself in the energy exchange while eating the banana.[514]

If your interest is improving your financial health, you might try the money diet. As you might guess, burn green candles in the kitchen and eat money attracting foods such as banana bread or banana cream pie. Fig newtons and grape juice are also good for this diet.[515]

Bananas supposedly have a correspondence with the planet Mars in promoting protection, courage, aggression, physical and magical strength and sexual energy.[516] Perhaps that last item is the reason why bananas are thought to have aphrodisiac powers. Of course, the banana's shape also has something to do with it. The ancients believed in the visual logic of the law of similarity. Simply put, foods that looked like genitalia were the best erotic stimulants. The Greek gods reflected this food and sex connection. Aphrodite, from whence cometh the term aphrodisiac, had a son, Priapus, the god of fertility. As such he was protector of garden produce, vineyards, flocks and bees. Artists represented him with fruit in his garment, sickle and cornucopia in hand, and a very prominent phallus. The ancient Greeks did not have bananas, but in Asia and Africa where bananas grow naturally, they are endowed with mythical properties associated with their shape.[517] The banana's shape, its soft, sweet flesh and creamy texture all contribute to its aphrodisiac reputation. The fact that bananas do contain substances which increase the sexually-related neurotransmitters dopamine and norepinephrine has something to do with it. But these are only in trace amounts leading one columnist to conclude that "I'm guessing by the time you consumed enough . . . bananas to elevate your dopamine level, you'd have bad breath and little inclination to move anything but your bowels anyway."[518] The whole aphrodisiac thing rests on shaky ground. How do you explain that in medieval Europe chicken soup was thought to be an aphrodisiac and in post-World War II Italy, Spam was the rage. It's all in the head anyway.

There may be many reasons to eat bananas, but the best reason why they eat them can be found in a poem from 1919.

Why They Eat 'Em

Some eat bananas if too thin
And some because they're stout;
Some say they eat because the skin
Keeps harmful germs without.

Some say the nourishment is more
Than found in chops or steak;
Some eat bananas by the score
For their digestions' sake.

Some vegetarians, scorning meat,
Find them a perfect food,
Some toothless one delight to eat
This fruit—so soft and good.

Some eat them at the breakfast hour,
To start the day aright;
Some, lunching, two or three devour
To last them until night.

Some have them at the evening meal—
They fill a long-felt want;
Some eat them anytime—they feel
They're not extravagant.

Some find them, when the weather's warm,
Refreshing as can be.
Some serve them, when the winters storm,
For greatest calorie.

Some cannot tell the reason why;
(It never seems to strike 'm)
ALL must admit—they *can't* deny
They eat 'em cause they LIKE 'M.
—"May I Knott"—[519]

CHAPTER 10

The Banana on Stage and Screen

"BANANAS is my business. You know
I make my money with BANANAS!"
Carmen Miranda

It is November 1926, and A. Robins is playing the Palace Theater in New York City. His biography in the theater program says he is Viennese and had been the imperial jester at the court of the czar of Russia, probably a position concocted by a creative publicist. Filling the third spot on the vaudeville program, a good position for an inventive clown, A. Robins appears on stage dressed in a Swiss Alpine outfit and carrying a fiddle and bow. Starting to play, he sees a spot on his tie and tries to brush it off. The whole tie disappears having been painted on the shirt. As he plays, the violin begins to shrink getting smaller and smaller until it almost vanishes. The music continues. The comedian produces all sorts of instruments from his coat—a trombone, a cornet, a bass violin, drums—and plays each one. Out come flowers, a music stand and campstool. When the music ends, the coat is traded for a new one which now produces endless numbers of bananas and generates gales of laughter from the audience. Magic? No, Robins' trick was that all the instruments had been collapsed to fill a small space and were spring released when pulled out.[520] The bananas? The coat had generous lining space. It was a sure fire vaudeville act.

Long before "Seinfeld" and Jay Leno, even before radio and movies, there was vaudeville, the live variety show that dominated American popular culture in the late nineteenth century and the first three decades of the twentieth. Vaudeville's roots can be found in American minstrel shows, English music hall, Yiddish theater, and West African music, humor and dance. Cyrus Townsend Brady, an Episcopal clergyman, described the program in a vaudeville theater on a steamy August night in 1901.

> Ladies in short skirts capered nimbly over the stage to the "lascivious pleasing" of the banjo, gentle tumblers frisked and frolicked about without the slightest regard either for temperature or gravitation; happy tramps—at least the announcements said they were happy—whose airy, carefully tattered garments were in full consonance with the heated atmosphere, delivered themselves of speeches full of rare old humor and fairly bristling with Boeotian witticisms. There were men singers and women singers, musical cranks, freak piano players, monologue artists, and then a little play—at least they said it was a play.[521]

Vaudeville appealed to a family audience; its close relative—some would say its actual father—was burlesque which appealed to a mostly male audience. Vaudeville featured music, comedy, sketches and short plays; burlesque became mostly musical acts and parody with heavily sexual overtones. Both entertainment forms evolved at the same time that the banana was becoming Americanized. Comedians became known as "bananas." Don Wilmeth in *The Language of Popular Entertainment* suggests that the use of the word may have derived from the banana shaped bladders that were used in burlesque for hitting the comedian over the head. Lead comedians were "first bananas." Others were "second bananas" and so on. The term "top banana" probably did not enter common usage until the 1951 Broadway production of *Top Banana* with Phil Silvers in the title role.[522] Nevertheless, it is clear that bananas, be they comedians or stage props, were always vaudeville and burlesque staples.

The B. F. Keith Vaudeville Circuit had a virtual monopoly in the eastern United States because of its ownership and booking of

Keith theaters and its booking of non-Keith theaters. The circuit was controlled by Edward Franklin Albee who also founded the dominating United Booking Office in 1906. Shortly thereafter he signed up an Australian swimming sensation by the name of Annette Kellerman, described in 1910 by one early commentator as "The Queen of Modern Vaudeville." Kellerman's career was promoted by her father who took her on a swimming tour of Europe where she swam down the Seine, the Rhine, the Danube and the Thames. She performed in an English music hall with an act in which she dived into a large, glass-enclosed tank and performed a water ballet. Albee signed her after she created a sensation by appearing in a one piece bathing costume at a Boston beach.[523]

The Keith-Albee circuit had a strict policy in its theaters against performers being obscene, vulgar or less than wholesome while performing. However, when Kellerman's act seemed to be a flop, Albee ordered mirrors set up all around the stage so that audiences could better view Kellerman in her daring bathing suit. As Albee told a house manager, "Don't you know that what we are selling here is backsides—that a hundred backsides are better than one?"[524]

Kellerman's act was often preceded by a freak act named "Blatz, the Human Fish." Blatz spent long periods of time underwater reading a newspaper, eating, and playing the trombone. Apparently, Kellerman liked the idea of doing more underwater than swimming, and she began eating a banana while underwater. Her act eventually became an extravaganza that employed two hundred "mermaids" at an enchanted waterfall. At other times she sang, danced and walked a tightrope along with the swimming.[525] No wonder she needed the banana—energy.

"Dr. Rockwell, Quack, Quack, Quack," was George L. Rockwell, one of the great "nut" acts of vaudeville. From the teens to the early 1940s Rockwell entertained his audiences by lecturing on human anatomy using a banana stalk to illustrate his points. Since bananas were very popular at the time, there was no shortage of stalks. After a career at all the major houses including Radio City Music Hall, Rockwell retired to a farmhouse in Maine which he called "Slipshod Manor."

If the name of this famous "nut" act rings a bell, it might be because of Rockwell's infamous son, George Lincoln Rockwell, the slain American Nazi Party leader.[526]

Many people might blame a banana peel for their downfall, but one burlesque comedian, "Sliding" Billy Watson, could say that the banana peel brought him fame and riches. "Sliding" Billy was born to show business. His father handed rabbits to Herman The Great, a magician. Billy played in medicine shows, carnivals. circuses and minstrel shows.[527] Then in 1900 he broke in with Fred Irwin's Burlesque Show.

"Sliding" Billy Watson explained in his own words how he discovered the slide which gave him his name and a vital part of his act.

> One day, years ago, I saw a fellow step on a banana peel just as he was going down a sharp incline in the sidewalk. At first his feet had a tendency to shoot from under him, but with acrobatic ability he managed to gain his equilibrium and he slid along. For a minute it was a good bet whether he would get the best of the struggle or whether the banana peel would accomplish its purpose. At last he got himself under control, but, even then, one couldn't tell for a second whether he was going forward or going backward on his head. It was the funniest stunt I ever saw and it made me laugh. Even then I was a hard guy to make laugh. The thought came to me that if a fellow could reproduce that slide on the stage he would be made, and I decided to experiment. I can tell you it took a great deal of rehearsing and I had to change the action and modify it in some respects, but at last I got the slide down where I wanted it. Then I called in some of my critical friends to see me do it, and they didn't have to say a word—whenI saw how they laughed I knew I was a winner. Since then the slide and I have been inseparable. I would no more think of cutting it out of my performance than I would of cutting off my right hand and saying that I didn't need it any more. And, believe me, I never go past a banana peel on

> the sidewalk now without feeling inclined to take off my hat and bow to it in a spirit of reverence.[528]

"Sliding" Billy Watson appeared on stage as a "Flying Dutchman" with red nose, chin whiskers, enormous shoes, baggy pants, a carrot wig and a vest split down the back. By 1921 he was successful enough to get his own troupe which included "a little scrawny Jewish kid named Fanny Brice" in the chorus.

Unfortunately, his stage success did not carry over to domestic life. In 1926 his wife, Mrs. Anna Shapiro ("Sliding" Billy's real name was William Shapiro), from whom he had been separated for twenty-four years, went to court and told the judge she was "working at the washboard" to support her children while her husband was "sliding around the country." The newspapers of the day had great fun playing off Billy's stage name. They reported that the judge promised "to slide him right into jail unless he kept abreast of the $7.50 per week support order. The judge permitted him to slide out of court without punishment. For at least the next five years "Sliding" Billy slid in and out of court with his wife in hot pursuit. In 1931 he managed to be rescued from a hotel fire in Philadelphia and slide into jail all on the same day. His stay was short, and he continued to slide around the country with fewer and fewer engagements as vaudeville and burlesque were losing the battle with the movies.[529]

Vaudeville and burlesque did not succumb quietly. They went with a banana wiggle and a lot of banana comedy. "The Banana Wiggle," a combo dance of the Charleston, the jitter-bug and the samba, could be seen at Minsky's Burlesque Theater at State and Van Buren Streets in Chicago in 1928 where the bill featured the lovely, curvaceous La Bomba, "The Human Heat Wave." What she waved, among other things, were bananas worn around her waist. A few years later at other Minsky theaters you might have caught Phil Silvers in a banana routine. Picture the sketch with three guys with a stage Dutch or Jewish accent and attired in funny costumes with baggy pants and big shoes. The three comics are standing center stage, and one of them holds up two bananas.

First Man: I have three bananas here, and I will give you one.

Second Man: You only have two bananas.

First Man: I have three bananas. Look, I'll show you. (Holds up one banana in his right hand.) One banana have I. (Holds up one in his left hand.) Two bananas have I now. One banana and two bananas makes three bananas.

Second Man: I only see two bananas. In your own words, I will show you you are wrong. (He takes the bananas.) One banana have I. Two bananas have I. One banana and two bananas . . . by golly! He's right! (Aside to the third man.) Would you like a banana?

Third Man (nodding): Okay. One banana for you. A banana for me.

First Man: How about me?

Second Man: You eat the third banana!

Black Out[530]

Two twentieth century women of the stage and screen gained a major share of their fame from association with bananas. Both rose from humble origins, and both left the land of their youth to achieve their success. Neither woman was ever completely comfortable with the reception she received in her homeland when returning as a recognized star. Josephine Baker left her St. Louis roots to gain fame on the French stage and forever was identified with her scanty banana costume. Carmen Miranda, the "lady in the tutti-frutti hat," left Brazil for Broadway and eventually Hollywood where she could accurately say, "bananas are my business."

Josephine Baker was born on June 3, 1906 as the illegitimate child of Carrie McDonald and Eddie Carson.[531] Carrie was a domestic and the descendant of Black and Apalachee Indian slaves. Eddie was a drummer in St. Louis entertainment houses of all types. At the time St. Louis was the capital of "Ragtime," the music popularized by its foremost composer, Scott Joplin. Carrie and Eddie put together a song-and-dance routine which they performed for a short while in the vaudeville houses and bars. Josephine's parents never married, and her father drifted away but not before providing Josephine with a brother, Richard. The abandonment led Josephine to create fantasy fathers. Through her lifetime Josephine reinvented herself when need, inclination or circumstances required. Her mother married Arthur Martin who adopted Josephine and Richard without legal formalities. The marriage provided Josephine with two sisters, Margaret and Willie Mae. Life was a constant struggle against bedbugs, rats and hunger.

Later in life Josephine told the story of an early traumatic event that heightened her awareness of the gulf between black and white society in America. One day in July of 1917 when rumors spread that a white woman had been raped by blacks, East St. Louis was torn apart by white mobs that overran and burned out the black ghetto. Thirty-nine blacks and nine whites were killed in a day and night of rape, pillage and burning. Josephine's mother hustled the family out of the house after midnight to flee to the other side of the Mississippi River. A terrified eleven-year old Josephine rescued her "babies," two puppies she had found half dead in a garbage can.[532] Like many episodes in Josephine's life, it may not have happened that way. Josephine's adopted son claims the stories Josephine told of the event were those of other people. She was in St. Louis, safely across the river from the riots in East St. Louis.[533]

At age thirteen Josephine left her parents' house, got a job as a waitress and met and married her first husband Willie Wells. Possibly she was pregnant, but no child resulted from the union. They soon separated. Said Josephine, "I hardly stayed with the man overnight."[534]

The first dollars Josephine earned in entertainment was with a medicine man's decrepit wagon show. She danced between the "old fraud's spiels" and helped him sell Indian herb laxatives and "suspect rejuvenators." She danced for "picnics, fish-fries, revival meetings, anywhere and everywhere."[535] Josephine joined the Jones Family Band which played ragtime outside bars and pool halls and passed the hat. Her first stage appearance came when the traveling group the "Dixie Steppers" played a black vaudeville house in St. Louis. The group was down an act and hired the Jones Family band to fill in. Josephine was a small and scrawny thirteen-year-old. Her first appearance was an accidental comic sensation. She was to play Cupid in a love scene. Gliding on pulleys above two lovers, the wires became entangled, and Josephine, clowning all the while, dangled helplessly overhead much to the delight of the audience. The manager had the wisdom to keep her with the troupe.[536]

The Dixie Steppers and Josephine played the black vaudeville "T. O. B. A." circuit. Known to the performers as the "Tough on Black Asses" circuit, it was actually the Theater Owners Booking Association for blacks. Ethel Waters, the famous black vocalist described the typical audience. "Rugged individualists all, they did whatever they pleased while you were killing yourself on stage. They ran up and down the aisles, yelling greetings to friends and sometimes having fights. And they brought everything to eat from bananas to yesterday's pork chops."[537] Audiences were hard to hold and hard to please. The main attraction was often the drawing for a door prize which might be a live turkey, smoked ham or a gold tooth. In such a setting Josephine developed a comic presence. Crossed eyes, slapstick antics and a deliberate awkwardness became her standard routine. It was her real desire to be a vocal soloist singing the blues in a full length dress bathed in a center stage spotlight—just like Ethel Waters. When the "Dixie Steppers" troupe broke up in New Orleans, Josephine was left stranded in a hotel room with "Voodoo" Jones and his wife Doll sleeping three to a bed.[538]

Josephine headed north to Chicago then east making herself available for any show seeking "tantalizing tans," "sepia lovelies" or "hot chocolates" for the chorus line. At age fifteen Josephine sought a spot in the chorus line of Nobel Sissle and Eubie Blake's *Shuffle Along*, the season's hottest show in Philadelphia. She was turned down because of her age, but while there she met and married her second Willie, William Howard Baker. Josephine's mother-in-law was upset by the marriage feeling that Willie had "married down."

Josephine was a chorus girl and a shade darker than Willie whose job as a Pullman porter kept him away from Josephine much of the time.[539] *Shuffle Along* moved to New York and Josephine followed. The show brought blacks to Broadway for the first time. Josephine's pestering finally won her a spot at the end of the chorus line in the show's road company. She stole the show with her clowning; crossing her eyes and tripping over her own feet while the other girls kicked with precision. The audiences loved her.

Back in New York in 1924 Sissle and Blake gave her an important spot in their new show *Chocolate Dandies*. She was billed as "That Comedy Chorus Girl—Josephine Baker." The *New York Times* review said, "As a freak Terpsichorean artist, Josephine Baker, with her imitation of Ron Turpin's eyes, made quite a hit."[540] Some of the reaction to the show and Josephine's role illustrate the dilemma for black performers in America at that time. Whites only wanted black performers to perform stereotypical roles. Blacks were criticized for trying to be too white. If blacks supplied "fast dancing and Negroid humor," as one white critic suggested, they were reinforcing dumb stereotypes and killing their own artistic growth.[541] Josephine Baker's whole career may be viewed as a struggle between exploiting the old stereotype and breaking new ground; it was a career that was anything but conventional.

In the summer of 1925 Baker was recruited by Caroline Dudley for a black revue in Paris. Dudley found Josephine performing at the end of the chorus line at the Plantation Club in New York. "She stood out like an exclamation point!"[542] Dudley had arranged with André Daven, producer of the Theatre Champs-Elysées, to put together a revue of thirty performers. Its architecture and performances made Daven's theater a "temple of the modern." It had premiered Stravinsky's *Rite of Spring* in 1913. But by 1925 the theater was operating in the red and needed something innovative for the fickle Parisian audience. The Parisian psyche had been captured in 1925 by African art and the theme of negritude as a result of the African Pavilion at the Exposition des Arts Décoratifs. Cubist artist Fernand Léger advised Daven, "Give them Negroes. Only Negroes can excite Paris."[543] His advice was prophetic.

When the nineteen-year-old Josephine arrived at the train station in Paris as part of the cast of *La Revue Nègre*, she was a slim teenager, a lighter color than the others, "a mulatto one shade darker than honey, standing with a stork like awkwardness of her young years that nevertheless implied a special grace, an engaging presence."[544] Josephine's first appearance on stage was in a riverboat wharf scene to sing "Yes Sir, That's My Baby." Critic Pierre de Regnier described her entrance. She "walks with bended knees and looks like a boxing kangaroo . . . Is this a man? Is this a woman? Her lips are painted black, her skin is the color of a banana, her hair, already short, is stuck to her head as if made of caviar, her voice is high-pitched, she shakes continually, and her body slithers like a snake."[545]

The closing scene of *Revue Nègre* was unexpected and startling. A solitary clarinetist played against a Manhattan skyline backdrop. With a trumpet fanfare Josephine was borne on stage upside down on the back of Joe Alex, the company's giant. She was bare breasted and wore only a collar and girdle of feathers. One observer remembered only one pink feather between her legs. To a snake charmer melody Josephine slithered down the "length of her muscled transport onto all fours" and weaved and coiled around her partner's legs. The audience reaction was mixed, but the applause won out over the whistles—the Parisian equivalent of boos. As the music became more frenzied, Josephine twisted and shook at the accelerated beat of a hummingbird. The dance was a combination of various moves related to belly dance—the Shake and Shimmy and the Mess Around—that were popular with black jazz dancers in the 1920s.[546]

French writers who had seen a preview earlier had a week to prepare their assessments of the show that was seen as taking the modern in several sensational directions at once. Josephine was "modern yet also a 'primitive,' an exotic flourishing in a European hothouse . . . ."[547] To Regnier, Josephine was "the triumph of lubricity, a return to the manners of the childhood of man." Classic dance critic André Levinson described her as a sinuous idol that enslaves and incites mankind . . . the plastic sense of a race of sculptors come to life."[548] "Perhaps the best way to imagine how sexy she looked to people brought up on the waltz is to remember how and why Elvis Presley once seemed obscene. It was equally a matter of violating white conventions of movement."[549] Maria Jolas, the co-founder of an avant-garde magazine, refused to see the show and said about Josephine many years later, "She just wiggled her fanny and all the French fell in love with her." The prevailing sentiment among the American colony in Paris was that Josephine had triumphed.[550]

The reaction to Josephine Baker and *La Revue Nègre* was as much to do about racial attitudes of the time as it was about talent. Robert de Flers, a reviewer for the most important Paris daily newspaper, thought the show "A lamentable transatlantic exhibitionism which makes us revert to the ape in less time than it took us to descend from it."[551] Many Europeans thought that with few exceptions, such as Buffalo Bill and Mark Twain, America had no culture. Jazz, with its Black musicians, singers and dancers, presented a black face of American culture to Europeans in the 1920s.[552] To the extent that Paris adopted Josephine, "she became another exemplar of the poverty-to-riches legend and proof of the great capital's world-inclusive, assimilationist, ever-renewing spirit."[553]

Josephine accompanied the revue to Berlin where her dancing was called a "frenzied fertility rite set to the syncopated rhythm of jazz." In less than two months she became as celebrated as Marlene Dietrich, Lotte Lenya and the Great Garbo. Right-wing Germans saw Josephine as a threat to the Aryan ideal, and she was condemned in pamphlets as "untermenschen," subhuman.[554]

Even before *La Revue Negre* moved to Berlin, Josephine had signed a contract with the Folies Bergère, an event that opened a new door to banana fame. The Folies Bergère is listed today in guide books as "erotic spectacle." It was founded in 1869 and was the first Paris music hall. It was the equivalent of American vaudeville. Bare breasted women first appeared there in 1894 but they always wore a G-string or bikini pants. Despite the fact that the Prefect of Police had to preview each show, the Folies were fairly vulgar in the early part of the century. By the end of the First World War nude girls were the principal draw. Yves Montand, Edith Piaf and Maurice Chevalier all started in music halls like the Folies.[555] It was for this venue that Henri Varna, the director of Folies Bergère, hired Josephine for *La Folie Du Jour* in 1926.

For *La Folie Du Jour*, Josephine received star billing. Her name appeared in lights in letters twenty feet high above the marquee. The production had three hundred employees and cost half a million dollars. Some of the production's music came from Irving Berlin. Josephine's famous act began when she made her appearance in a jungle setting walking backwards on hands and feet down a tree trunk. She wore only "the ultimate costume of erotic suggestion," a girdle of rhinestone-studded plush bananas around her waist, While Josephine danced, the "bananas trembled like Jell-O on a fork." Taylor Gordon, a black American singer who saw the show, said Josephine "was flopping her bananas like cowtails in fly time."[556] It was much like her savage dance from *La Revue Négre* done in a jungle setting. Later in the program a large iron egg-shaped cage, painted gold, descended from the ceiling of the theater to stage level. The egg opened to reveal Josephine standing on a large mirror. Wearing a grass skirt and feathers around her neck, Josephine, as if possessed, danced the Charleston in a visual blur.[557]

Bananas, nature's witty phallic symbol, became Josephine's professional trademark although her adopted son believed "there was nothing prurient about all those swinging bananas, they were funny."[558] Josephine, "the Ebony Venus, the Black Pearl, the Creole Goddess," was reported by newspapers to be the most photographed girl in the world. She was interviewed and written about by every Parisian publication. Her picture on postcards outsold pictures of the Eiffel Tower. Alexander Calder sculpted her as a wire mobile. Cocktails, bathing suits and hair goo were named for her. Banana-clad Josephine dolls sold by the thousands, and by Christmas 1926 they were the rage in France and Harlem. In Paris they shared the Christmas windows of the Galeries Lafayette with Santa Claus and the Virgin Mary. Parisian designers gratefully clothed her gratis, and she brought extravagant costuming back into fashion.[559]

Josephine Baker in Paris in 1927 in one of several versions of her banana belt. (Courtesy New York Public Library, Billy Rose Theatre Collection)

Josephine's bananas were the most talked about and cartooned costume in Europe. Where did the costume idea originate? Some claim it was the inspiration of Paul Poiret, the couturier. Josephine claimed that the writer Jean Cocteau was responsible. "It is Cocteau who gave me the idea for the banana belt. He said, 'On you, it will look very dressy.'"[560] Whatever the inspiration, Josephine's bananas had profound effects. A film clip of Josephine's banana dance, with breasts covered for an American audience, was shown in the United States. A spectator in Columbia, South Carolina became so excited he sprang to his feet and yelled louder and louder, "Do Jesus! Do Jesus! I have never seen a woman dance so good before in my life."[561]

Josephine was the toast of Paris and was often seen in the clubs of Montparnasse, an unsavory Paris pleasure district. Ernest Hemingway met her there at "Le Jockey" where they danced virtually nonstop. Josephine never removed the black fur coat she was wearing, and at the end of the evening she surprised her dance partner when she said she had nothing on underneath.[562] At home Josephine was known to sometimes greet visitors wearing nothing more than a pleasant expression.

Josephine's success attracted the attention of Guiseppe Abatino or "Pepito" as he was known. Actually a simple bricklayer and plasterer, he assumed the title of Count—a "no account count" in the words of one of Josephine's friends. Pepito handled her money, contracts and correspondence and was her lover when she needed him. An announced wedding was a sham.

Josephine's 1927 season at the Folies, her second, was a newly staged show. She again wore a banana skirt in a more "spangled hard-edged version." "It was the fate of those bananas to become ever harder and more threatening with the years, so that at last they looked like spikes." Josephine was as striking and dancing as well as ever, but bare breasts and bananas aroused little excitement with a fickle public. The glitter began to fade.[563] That same year Josephine co-authored the first of four autobiographies; she reinvented her background often enough that one would never do. She also offered beauty tips to Josephine wannabes. "The best toilet water is rainwater, it keeps indefinitely. Bathe in the milk of violet petals. Moisturizers made from bananas fight wrinkles. Rub strawberries on your cheeks to give them color. Swim as often as you can. Animals which live on land are never as elegant as fish."[564]

Josephine made a disastrous first film appearance in 1927 in the silent *La Sirène des Tropiques*. Pepito played her husband in this ludicrous tale of a tropical girl who becomes a successful dancer in Paris and fends off the threatened rape by a great white hunter. It was time for a change, and Pepito arranged for a world tour of twenty-five countries that began in Vienna where posters showed Baker wearing only her bananas, a scene more shocking in Vienna than in Paris or Berlin. Catholic Vienna challenged the presence of the "Congo Savage." Josephine calmed the storm somewhat when she opened the revue in a modest floor length evening dress singing "Pretty Little Baby" demonstrating that she was not the devil's mistress.[565]

The tour moved on to Budapest where angry students attacked the audience with ammonia bombs and screamed "Go back to Africa." In Munich her performance was forbidden by the police on that grounds that it would cause public disturbances and corrupt morals, but her race may have been more of a factor than her dancing. In Bucharest she performed in an open-air theater. Rain started when the show began, and it was pouring when Josephine appeared. She abandoned an umbrella. Her bananas got soggy, dropped off and plopped on the stage. After the European tour and shaken by the racial hatred she had experienced, Josephine set out for South America.[566]

Josephine played to full houses in Argentina, Chile and Brazil. A riot triggered by internal politics erupted in Buenos Aires. Controversy and disruption seemed to be almost part of the act. Returning to France Josephine had a comic encounter aboard ship with LeCourbusier, the architect, who was returning from a lecture tour. At a "crossing-the-equator" masquerade ball the famed architect appeared with blackened skin and wearing a feathered codpiece imitating Josephine's Folies costume. The entertainer and the architect paired off for the remainder of the voyage.[567]

On her return to Paris in April 1929 Josephine told newspaper editors "I don't want to live without Paris. It's my country. The Charleston, the bananas, finished. Understand? I have to be worthy of Paris. I want to become an artist." She respectfully packed her banana skirt in a trunk. Later Pepito took it out one day and put it on. An indignant Josephine said, "You shouldn't make fun of the tools of a person's trade."[568] But despite her intentions, the bananas never quite disappeared from Josephine's life.

In 1930 Josephine appeared in *Paris Qui Remúe*, a new and relatively decorous revue, at the Casino de Paris. As a "lake fairy" she wore a jewel-studded G-string created by Van Cleef and Arpels. Josephine captured Parisian hearts with the song that became her trademark, "Jai Deux Amours." In a poster for the show Josephine appeared with a baby leopard, "Chiquita"—yes, Chiquita—given to her by producer Henri Varna. "Chiquita" went everywhere with Josephine including a performance of *La Boheme* at l'Opéra where during Mimi's death scene Josephine "accidentally" released "Chiquita's" leash. The leopard ended up in the orchestra pit as Mimi, and probably a large part of the audience, expired.[569]

Josephine was often referred to in animal terms; "monkey" was one of them. She loved animals and said she learned to dance by watching animals in the St. Louis zoo. Her pets included a monkey named Ethel, rabbits, macaws, goldfish, a house-trained piglet and a boa constrictor that sometimes doubled as a neck piece.[570]

In keeping with her star status, Josephine maintained a country home, Beau-Chène, north of Paris where she engaged in gardening and showed a motherly concern for the neighborhood children. The neighbors also saw her wading playfully through the waterlilies of the duck pond while "splendidly nude."[571]

Josephine made another film appearance in a 1934 light, romantic comedy, *Zou Zou*. Pepito promoted the film with an advertising campaign that revived Josephine's banana associations. Fruit merchants were asked to conspicuously display mounds of bananas with stickers saying "Josephine Baker is Zou Zou." He paid doctors to submit articles to Paris newspapers endorsing the nutritional aspects of bananas. He also tried to convince cinema owners to dress ushers in banana skin outfits, but none did. The film was not memorable, and the banana promotion flopped.[572]

Josephine returned to the United States to appear in the 1936 edition of the *Ziegfeld Follies* and discovered that America would never welcome her the way France did nor be able to look at her without considering her race. Fanny Brice got exclusive headline billing with Bob Hope and Josephine Baker lower on the marquee. Josephine had three appearances in the show including a torrid jungle scene where she wore a throwback to the banana belt. This time the costume designed by Vincente Minelli was a waistband of aggressively pointed silver tusks. The critics did not treat her well. Brooks Atkinson of the *Times* said Josephine "has become a celebrity who offers her presence instead of her talent," and "she has refined her art until there is nothing left of it."[573] *Time* magazine in a racist essay wrote that her success was undeserved, and "in sex appeal to jaded Europeans of the jazz-loving type, a Negro wench always has a head start." Josephine raged at Pepito, and he returned alone to France where he died of kidney cancer. On her own Josephine opened a nightclub New York in the east fifties where she enjoyed some success before returning to Paris.[574]

When World War II broke out in Europe, Josephine joined Maurice Chevalier in performing for French troops at the front lines. Before the Germans invaded the country, Josephine appeared in a new review, *Paris-London,* promoting French-British friendship. She worked for the Red Cross and wrote hundreds of letters and sent gifts to soldiers at the front. When the Germans overran France and set up the puppet Vichy French government in the south, Josephine declared she would never sing in Paris again "as long as there's a German in France."[575] To add insult to injury, Hermann Goering took over Josephine's Paris townhouse and later invited her to dinner in her own house. Allegedly, the invitation was part of a plot to poison her, but Josephine managed to survive the episode.[576]

In the Spring of 1941 Josephine was in neutral Spain where she had not appeared for ten years. At appearances in Madrid, Valencia and Barcelona audiences were surprised that she was not still wearing bananas. She said she had had to eat them because of the war. Because of her celebrity status many doors were still open to her, and she used the opportunity to collect intelligence for the French resistance both in Spain and Vichy France. She sometimes crossed borders with scraps of paper in her brassiere and panties. After the war the French recognized her contribution with the Légion d'Honneur and Médaille de la Résistance.[577]

After the war in Europe ended Josephine waged another war in her homeland, this one for civil rights. In 1948 she toured the United States with her new husband Jo Bouillon and his orchestra and experienced the harsh reality of segregation. She was refused reservations at thirty-six New York City hotels. She traveled incognito though the South. When she was thrown out of white lunch rooms and ladies rooms, she went back in.

Back in France she decided to return to the United States to help her people. For a 1950 American tour she required that black stagehands and musicians must be hired for every show where she appeared. She only performed in cities where she received accommodations in first class hotels. In Miami Josephine was the first African-American to perform before a mixed audience and to become a resident in a white Miami Beach hotel. She promoted civil rights with business leaders in Chicago, San Francisco and Los Angeles. She was threatened by the KKK, but her tour was a professional success. "Her political views were a mixture of naiveté, a messianic will to make men love one another, and a desire to get back at America for not accepting her as the instrument of its deliverance from racism.[578] Josephine appeared at the 1963 "March on Washington" where commentators said she and Martin Luther King, Jr. were the most hopeful speakers of the day. With optimism she said "you are on the eve of a complete victory. You can't go wrong. The world is behind you," and added "salt and pepper, just what it should be." For Josephine, the March was the end of her civil rights struggle in America.[579]

Josephine's idea of world brotherhood was put into practice in 1953 when she launched her plan for a "Rainbow Tribe." She wrote to an orphanage in Japan, "I would like you to find me a Japanese boy of pure race, a healthy one, two-years old. I want to adopt five little two-year old boys; a Japanese, a black from South Africa, an Indian from Peru, a Nordic child, an Israelite; they will live together like brothers."[580]

"The Rainbow Tribe" eventually numbered twelve orphans of various hues and nationalities. It became an increasing burden to support both the children and Josephine's showplace French country chateau, "Les Milandes." When Josephine appeared in Berlin in 1968 she told her audience, "Oohlala, I'm not ready, in the old days it was easier, only a few bananas, but today my children have eaten those bananas." And back at her estate times were so bad she could not buy bananas for the monkeys which sometimes numbered thirty.[581] In 1969, despite a TV fund raising appeal by Brigitte Bardot, Josephine and the "Rainbow Tribe" were evicted from a foreclosed "Les Milandes." Princess Grace of Monaco provided a Monte Carlo villa for Josephine and her brood.

Josephine was still making stage appearances. In 1973 she performed for four nights at Carnegie Hall to favorable reviews. In 1974 she appeared at the Palladium in London to help needy actors. With the Queen Mother in attendance, Josephine told old stories. "I started in 1924, and we were all beginners together—Pablo [Picasso], Matisse, Hemingway. I used to look after them, picking up their clothes, getting them organized. As for the bananas, I wasn't really naked, I simply didn't have any clothes on."[582]

On April 8, 1975, exactly fifty years after her semi-nude appearance in *La Revue Nègre*, she made her final triumphant appearance in *Josephine* at the Bobino Theater. The morning after her second performance, Josephine was found in bed in a coma after suffering a cerebral hemorrhage. With Princess Grace at her side, Josephine died in Paris at the Salpêtrière Hospital. The funeral was held at the Church of the Madeleine built by Napoleon. Twenty thousand Parisians lined the funeral route to see the tricolor draped casket and hear the twenty-one gun salute. The hearse passed all the main landmarks in Josephine's Paris career including the theater of her current show where "Josephine" blazed in lights. Today at the Josephine Baker Museum in Dordogne, which is filled with waxen figures depicting scenes from Josephine's life, you can see the banana costume "exhibited with panoply worthy of Nefertiti's right elbow."[583]

Examine television film listings almost any week of the year and you are likely to find *The Gang's All Here*, the 1943 film in which Carmen Miranda performed "The Lady in the Tutti-Frutti Hat" and which forever identified her with the banana. The Miranda image has been frequently parodied by actors in drag including Bob Hope, Jerry Lewis, Mickey Rooney, Bugs Bunny, Willard Scott, and Ken Russell. At a 1991 gala benefit for the Rain Forest Foundation it was rumored that Elton John would appear dressed as Miranda, but it didn't happen.[584]

Sometimes the image is the wrong one. In early 1996 the *National Law Journal* wrote "Many legal experts believe the [mob hit man Harry] Aleman case . . . could go all the way to the U. S. Supreme Court. That could make Mr. Aleman's name on double jeopardy what Carmen Miranda's became on self-incrimination." The *Washington Post* noted that it was Ernesto Miranda who incriminated himself; Carmen Miranda established the right to pile fruit on your head.[585]

The "Brazilian Bombshell" was actually born in Portugal in 1909 as Maria do Carmo Miranda da Cunha, the second daughter of a barber father and an unschooled weaver mother. The family soon started calling her Carmen which along with Miranda, her mother's maiden name, became her stage name years later. Carmen's father departed for Brazil in 1910 looking for a better life. Carmen, her mother and sister soon followed. In the next five years two brothers and two sisters arrived in Carmen's family. They lived in a poor "red-light" district of Rio de Janiero. From 1919 to 1925 Carmen attended a Catholic convent school, but at age fourteen she had to give up formal education and began decorating windows and selling ties to elegant gentlemen at a haberdashery. In late 1925 the da Cunha family opened a boarding house in the business and market section of Rio. It was a move that opened new opportunities for Carmen.[586]

As a fifteen-year-old, Carmen became an apprentice milliner

and worked in the family boarding house where she often met musicians rooming there. In 1928 she was "discovered" by Josué de Barros, a composer and guitarist. At the time Carmen was a "five-foot-one, slight and attractive young woman who reinterpreted songs adding energy, drollness and sensuality to their lyrics." Her voice was unpolished, but she added a spontaneous "Brazilianness" to popular themes.[587] With Josué's guidance Carmen recorded two of his songs in 1929. The following year Josué learned that RCA Victor was interested in developing a collection of popular Brazilian tunes. At his suggestion she auditioned and won a one year contract. In the first year she recorded forty different titles. In the next decade in Rio she recorded 281 tunes. Half were sambas and the other half were Carnival marches.[588]

Between 1930 and 1939 Carmen was active in Brazil's musical world and became one of the principal voices on radio. Audiences were aware that her songs originated with the common people of Brazil, and they packed the auditoriums where her broadcasts were done. She began to develop a public image not unlike that of Eva Peron in Argentina some years later. She dressed in expensive and striking fashions, and women of all classes copied her outfits. The common folk expected her to outshine the Brazilian aristocracy.[589]

Carmen appeared in six full-length films in Brazil in the 1930s. In the last of these, *Banana da Terra*, filmed in 1938, Carmen changed her image and style. She adopted the costume, seductive gestures and tropical gaiety of the Bahian image which she extracted from a description in a song, "O que é qué a bahiana tem?" (What does the Bahian girl have?). The baiana costume was typical of black women who cooked and sold delicacies while they sat on doorsteps in the city of Bahia. But Carmen's baiana costume and expressive dance were almost beyond description, although many tried. Robert Sullivan, a New York writer, in 1941 said, "dramatic critics have grown neurotic in their attempts to get into words that would make sense and at the same time not brand them as mad sex fiends [what it is that Carmen Miranda has]." He noted that her costume covered her except for the space between the seventh rib and a point at about the waistline. "This expanse is known as the 'Torrid Zone.' It does not move but gives off invisible emanation of Roentgen rays." To add height to her stature, Carmen added a towering flamboyant headdress with two little baskets of fruit. The style was completed by Carmen's more than expressive body movements about which one observer remarked, "Carmen Miranda can't sing sitting down, because when she sings every muscle, every fibre of her anatomy moves and quivers."[590]

This was the Carmen Miranda, the "Queen of Samba," that American producer Lee Shubert wanted to see on a visit to Rio in 1939. Carmen had an obsession for many years to go to the United States and readily agreed to a contract to go to Broadway to appear in *The Streets of Paris*. Accompanied by her own band, "Bando da Lua," Carmen departed for Broadway on May 4, 1939 aboard the S. S. *Uruguay*.

When Carmen arrived in the 1939 World's Fair city of New York on May 17, she was met by a sizable contingent from the press, the Shubert organization and the Brazilian Consulate. The *World Telegram* reported that before any reporter could ask a question, Carmen blurted out, "I say money, money, money, I say twenty words in English. I say money, money, money and I say hot dog! I say yes, no and I say money, money, money and I say turkey sandwich and I say grape juice."

The *New York Post* reported she knew ten English words and said "monnee" (sic) five times.[591] Either way Carmen knew where American priorities lay.

In the *Streets of Paris* Carmen's songs were not listed in the program, and she appeared for only about six minutes at the end of the first act. But she was seen as the "outstanding hit of the show." Brooks Atkinson of the *Times* said, "she radiates heat that will tax the Broadhurst [Theater] air-conditioning plant." In a later review in which Atkinson noted that Carmen sang in Portuguese rather than in Spanish as he earlier reported, the reviewer said, "she sings with an incandescence of personality that enkindled the first night audience the instant she appeared."[592] The audiences did not understand a word that Carmen was singing; they didn't need to. It was the energy, fun and magic that Carmen exuded that they appreciated.

In 1940 the show went on the road, and in March the Brazilian Bombshell got a chance to perform for Eleanor and Franklin Roosevelt and their guests in the White House where thoughts were temporarily diverted from the real bombshells then dropping in a war torn Europe. For the occasion the "Torrid Zone" was covered by a white brocade dress, but fruit still decorated an immense turban.

Even before her White House performance, Carmen was destined to play a role with political overtones. Darryl F. Zanuck, vice-president of Twentieth Century Fox, had decided in 1939 to promote Roosevelt's "Good Neighbor Policy" and his studio's Latin American market with what became known as a new genre of film—the "banana movie." *Down Argentine Way* was the first of the genre to mix Pan-Americanism and Latin American dance including Carmen's samba. The film starred Don Ameche, Betty Grable and introduced Carmen Miranda to filmgoers. The film was shot on sets in Hollywood and New York. All of Carmen's scenes were shot in New York, making her a Hollywood star in absentia. In his book *Hollywood Musicals*, Ted Sennet said the film "is of no consequence, but as empty-headed nonsense, it is diverting." Of Carmen he said, "Her arms waving, her eyes rolling, and her hips swaying in rhythm with the music of her own Banda da Lua, she sings 'South ['Souse'] American Way' with an energy that threatens to topple the absurdly tall and elaborate hat [worth $300] from her head."[593] The film was a success in the United States, but it so distorted the reality of South America that it was banned in Argentina and infuriated many in other Latin countries.

After about a year in the United States, Carmen returned to Brazil in July 1940. The *New York Times* reported that she received one of the greatest ovations that Rio had ever witnessed, but when she appeared in a benefit performance at Cassino da Urca and sang some of the songs she had recorded in New York, she was received very coolly by a wealthy audience that resented the "negative and ridiculous" image she was projecting to the world. After a two month rest she planned a comeback. With new songs she returned to Cassino da Urca and gave what a Rio newspaper described as one of her best performances ever. With a new Twentieth Century Fox film contract in hand and a plan with Lee Shubert for a five week theater engagement, Carmen went back to New York in October 1940 aboard the S. S. *Brasil* to take the Bahiana to Hollywood.

With the outbreak of war in Europe, Nelson Rockefeller became concerned about hemispheric defense and strengthening the bonds between North and South America. In 1940 he persuaded Roosevelt to create an Office of the Coordinator of Inter-American Affairs (CIAA) to promote "good neighborism" in every possible way through radio, film, press and cultural exchange. John Hay Whitney was placed in charge of the Motion Picture Section of the CIAA. For about five years this office stubbornly insisted in putting an American spin on everything related to Latin America. For example, Disney's *Three Caballeros* was produced as powerful propaganda to promote democracy. Whitney's Section also advised MGM to highlight the character played by Dezi Arnez in *Bataan*. Not surprisingly, Carmen Miranda was to play a large role in the efforts of the CIAA.[594]

That Night In Rio was Carmen's next "good neighbor" film and starred Don Ameche and Alice Faye. Carmen had the first and last numbers in the film. They included some of her best known songs, a rhumba, "Chica, Chica, Boom Chic" and a conga, "I, Yi, Yi, Yi, Yi, I Like You Very Much." Neither was a samba. During the filming, Darryl F. Zanuck offered the director advice on how to best handle Carmen's on screen appearances. In a memo he wrote, "Your handling of Carmen [Miranda] was good except that she does not need to talk nearly as much as she does in English. You should confine her English to one or two line speeches, and if she has to explode, she should go into Portuguese—ending up with a denunciation of one or two words in English." He also ordered, "you cannot cut away from Miranda once she starts to sing."[595] It was a recommendation that many were to follow. In Brazil, *That Night in Rio* was considered Miranda's best film.

Weekend in Havana was Miranda's next film. The feature number was "The Nango" which Hermes Pan described as "Two hundred gorgeous feminine Zombies and Carmen Miranda in a voodoo jive."[596] When it opened in 1941 its first week box office figure of $25,000 soared past *Citizen Kane's* second week's gross of $9,000. In Cuba it was a different story. Most Cubans found the film grotesque and regarded Miranda's character of "Rosita Rivas" as absurd.[597] So much for promoting hemispheric relations.

While Carmen was pained by negative reactions in Latin America, particularly in Brazil, she was a complete success in the United States. By April 1942 with less than a year and a half in films, Miranda had more commercial connections than any other star in Hollywood. A studio announcement proclaimed she was involved in publicity for "furs, cosmetics, radio, coffee, dresses, hats, over a dozen sorts of games, books, phonograph records and a number of other items, including a bathing suit." In 1940 she sat for her portrait by New York artist Paul Meltsner who was known for his paintings of Gertrude Lawrence and Martha Graham. Carmen's portrait showed her in a pensive mood with a basket of green bananas and a guitar in the background. In March 1941 Carmen became the first South American in *Variety's* words "to plant her tootsie in a square of sloppy concrete" at Grauman's Chinese Theatre.[598]

Miranda returned to Broadway to appear in *Sons O' Fun* which opened in New York at the same time that Japanese bombs were dropping on Pearl Harbor. The show was a bit crazy but far from a bomb; it grossed $41,000 in its first week. Carmen's next movie, *Springtime in the Rockies*, was a box office leader in 1942. Carmen romped through the film playing Rosita Murphy, a half-Irish, half-Brazilian girl. But it was her next film that made her the Queen of the "Banana Movies."

Busby Berkeley did only one film at Twentieth Century Fox. It was fortuitous for film fans that Berkeley and Miranda collaborated on the 1943 film *The Gang's All Here*. The Fox musicals of the 1940s were "frothy, frivolous and flashy" with an emphasis on big band swing music which in the case of *The Gang's All Here* was provided by Benny Goodman. Berkeley, doing his first color film, continued that Fox tradition. Martin Rubin has written that the film was "a kind of macro-production number from beginning to end." Its "narrative and [production] numbers exist on a precipitous level of garish delirium." It exemplified Berkeley's reputation as the "field marshall of massed armies of screen dancers." Rubin believes that Berkeley lacked a concept of film as a whole resulting in the imposition of a spectacle style across the entire film. The story line was typically trivial with a blandly contrived romantic misunderstanding. The verbal level of the film, both script and dialogue, were "largely reduced to level of meaningless babel."[599] In short, *The Gang's All Here* is sheer spectacle—and what a spectacle.

The Gang's All Here did its part for the Good Neighbor policy. In the film the S. S. *Brazil* unloads the major exports of Latin America: sugar, coffee, fruit and Carmen Miranda. In a nightclub scene the host comments "Well there's your good neighbor policy," and Carmen teachers the audience the "Uncle Sam-ba." But it is "The Lady in the Tutti Frutti Hat" production number that is most memorable because of its banana motif.

It is reported that Ivonne Wood, the costume designer did not like mixed-fruit attire. Thus she settled on a headdress for Carmen using only bananas and strawberries. From there Berkeley's genius took over with strawberries and bananas everywhere. Carmen enters the scene sitting on a bed of bananas in an ox-cart. With Berkeley's single camera zooming around the scene, a battalion of scantily-clad girls cavort in an Afro-Equatorial setting with what have become legendary five foot papier-mâché bananas. They manipulate the bananas into a series of suggestive formations. From overhead the camera sees geometric patterns with the giant bananas and giant strawberries and spread-eagled chorus girls.[600] In the final shot Carmen's headgear reaches to banana infinity.

In *Agee on Film* the author writes about the banana scene and the camera "cutting to thighs, then feet, then rows of toes, which deserves to survive in every case-book of blatant film surreptition for the next century." The *New York Times* reviewer also caught the innuendo. "Mr. Berkeley has some sly notions under his busby. One or two of his dance spectacles seem to stem straight from Freud and, if interpreted, might bring a rosy blush to several cheeks in the Hays Office."[601] A more recent view of the matter said "The Lady in the Tutti Frutti Hat" number in *The Gang's All Here* "lampoons both United States-Latin American trade relations and notions of feminine sexuality through the casting of Miranda as the overseer of countless enormous, swaying phallic bananas buoyed up by lines of chorus girls who dance above other girls who have oversized strawberries between their legs."[602]

Whatever the interpretation, the film had a lasting impact. It inspired an immediate Latin influence in fashion that lasted through the 1940's, and today it remains a camp favorite. In 1978 *The Gang's All Here* had an eight week engagement in New York and a six week run in San Francisco. The theater lobbies were decorated with enormous plastic bananas, and fifteen hundred of the real thing were given away to opening day patrons.

In 1944 the *New York Herald Tribune* noted that in her pictures "Miss Miranda has worn wooden chopping bowls, bananas, palms, mixed green salads, Christmas toys and fruit compotes on her head." There was money in garish headgear. In 1945 Miranda with a salary over $200,000 was listed ninth and the highest ranking woman on a list of stars' incomes. She was ahead of Bing Crosby, Paulette Goddard, Bob Hope, Errol Flynn, Cary Grant, Betty Grable, Dorothy Lamour, James Cagney and Humphrey Bogart. But fame can be fleeting. *Something for the Boys* did well at the box office, but *Greenwich Village* was a financial flop and *Doll Face* was mutilated by the censors. In *If I'm Lucky* Bosley Crowther of the *Times* called her an "animated noise." By 1946 Miranda was not among the top thirty grossing Hollywood stars.[603] While her career took many ups and downs, the bananas remained a constant.

Carmen was becoming bored with always playing a banana character, but those around her were convinced that the best thing to do was to keep exploiting the banana image. So it was not surprising that Carmen would comment to a nightclub audience, "But you know something . . . BANANAS is my business. You

Carmen Miranda, in fruit bowl headgear—one of many variations. (Courtesy New York Public Library, Billy Rose Theatre Collection)

know I make my money with BANANAS!" And so she did be it television or nightclubs. When she appeared on Milton Berle's Texaco Star Theatre, Berle came out in drag as Miranda's sister and took a banana from Carmen's headdress and ate it—a sight gag used by other TV hosts. "I Make My Money With Bananas" from Oscar-winning composer Ray Gilbert became part of her nightclub performances from Europe to the United States.[604]

It was a tired and ill Carmen Miranda who appeared at the NBC Studios to perform on the Jimmy Durante Show in August 1955. With no general audience for the taped show, Carmen performed "Delicado," a romantic lament that spoke of "pain here in my heart." During the second half of the show, Carmen suffered a fall. Out of breath, she collapsed momentarily. Jimmy helped her to her feet, and the show went on. At home after the show Carmen sang, danced and celebrated with out-of-town guests. She went to her bedroom about 2:30 am. It was there the next morning that her husband, David Sebastian, found her body on the floor. Her makeup of the night before was still intact.[605] She was forty-six.

Maria do Carmo Miranda da Cunha was buried in Rio de Janiero, but the Carmen Miranda legend lived on. Carmen's final appearance on the Jimmy Durante show was telecast several weeks after her death. Durante and NBC were reluctant to show it, but NBC reported that Carmen's family requested it be used.[606]

Carmen Miranda wannabees became as common as bananas. There was Lina Romay, "Cugie's Latin Doll," at MGM, Margo of RKO, Maria Montez at Universal, Acquanetta, the "Venezuelan Volcano," and Olgo Jan Juan, the "Puerto Rican Pepperpot." It has been suggested that the United Fruit Company's "Chiquita Banana" character was inspired by Carmen Miranda. Two young songwriters supposedly wrote "Carmen Banana" for Carmen Miranda, but she rejected it. Later when they were working for a New York advertising agency they changed the title to "Chiquita Banana" which is the title used when the song appeared in 1944 at the height of Carmen Miranda's career.[607] It is a bit sad that Carmen Miranda's crown of bananas became a creative straitjacket, but it is her vitality and the fruit headdress that will keep her films on television for years to come.

While Carmen Miranda's career was slowing down, that of Phil Silvers was about to peak. Silvers was a first banana of burlesque in the 1930s before he moved on to theater and then to Hollywood. In 1951 he became the star of the burlesque-inspired musical *Top Banana*, the show that actually coined the term for a lead comedian. The show opened at the Winter Garden Theater in New York on November 1, 1951. Silvers played the role of "Jerry Riffle" which was reputedly modeled on Milton Berle. Berle himself played the role in summer theater in 1963. *Top Banana* ran for 356 performances before starting a national tour. The show became a film in 1954. The film, shot in only six days, was almost a direct shoot of the stage show.[608]

Brooks Atkinson, the *New York Times* critic, noted that "the score for *Top Banana* is negligible—hardly more than a reflex action to the necessity of having something that passes for music in a musical show." *Variety* in the film review said, "The plot would fit a thimble." What the show did have, in the words of Atkinson, was "exuberant performers. . . . Phil Silvers and a sociable crew of burlesque mountebanks careen through the evening with so much gusto that their show is likely to remain the most hilarious in town for quite a long time. For Mr. Silvers' barrage of gags and routines is uproarious—wild, helter-skelter and bouncing. . . . A theatergoer with a frail constitution should be cautious about attending *Top Banana* without a certificate from his physician."[609] Because Silvers worked so hard at the old burlesque hokum, Atkinson expressed concerned that the stage buffoon would become a physical wreck.

The role of "Pinky" in *Top Banana* was played by Joey Faye, one of the greatest "second bananas" whose career spread over nearly seventy years. His real name was Joseph Palladino, the son of Italian immigrant parents. In his career he experienced a wide range of entertainment media. He was a fellow star in burlesque with Gypsy Rose Lee at Minsky's. She was famous for her striptease act, but she never took all of her clothes off. Fiorello LaGuardia, the mayor of New York, would not allow such behavior for fear that the sight of nude bodies would corrupt young people. Faye once recalled that "rough" language such as "hell" and "damn" were not permitted on stage either. As burlesque declined, Faye moved on to Broadway where he appeared in thirty-six shows and to Hollywood where he provided comic diversion to film stars such as Cary Grant, John Wayne and Gary Cooper.[610] In the 1980s near the end of his career he appeared as a bunch of dancing grapes in Fruit of the Loom commercials. Too bad he couldn't have been a banana.

No discussion of bananas on stage and screen is complete without mention of Woody Allen's 1971 film *Bananas*. Someone once asked Allen why he named the film *Bananas*. In typical Allen fashion he said the title was because there were no bananas in the film. He was right. There were no bananas, but there was a banana republic, San Marcos—actually Puerto Rico. Fielding Mellish (Woody Allen) is the little man in a comic struggle with machines, food, and women in New York and the tropics. The film "parodies the Cuban Revolution, American interventionism, Black Studies, the CIA, TV advertising, the military Right, the loony Left, machines, profits and a number of sacred cows dear to Middle America while at the same time, spoofing such solemn revolutionary biopics as *Viva Zapata* and paying homage to the Marx Brothers and *Duck Soup*."[611]

The film was extensively improvised and often depended on suggestions from crew and cast. There was little if any rehearsal, and scenes were often shot over and over again until the dialogue, camera angles, and action were satisfactory. Many situations are improbable including the one which made *Bananas* one of the favorite films of horror author Stephen King. Said King, "I just love that moment where this huge black lady is on the [witness] stand and identifies herself as J. Edgar Hoover. When the prosecutor questions her about her appearance, she says, 'I have many enemies.'"[612] This courtroom scene, done as a cheaper alternative to a chase scene, takes place after Fielding (Allen) has become the revolutionary President of San Marcos and returned to the United States where he is put on trial as "a New York Jewish intellectual Communist crackpot." Fielding is convicted, but the judge suspends the sentence on condition that Fielding promise not to move into the judge's neighborhood. Without doubt *Bananas* is bananas.

In the early 1990s, if you lived with any wee people, you were likely to see B-1 and B-2 on your television screen. They are two identical twin bananas who wear blue and white-striped pyjamas and speak in Australian-accented men's voices. They should because the show, *Bananas in Pyjamas,* originated Down-Under in 1992. It entered the American market in 1995 and by 1996 was syndicated in thirty-two other countries including Saudi Arabia and Iceland. In 1997 the banana duo earned $6.7 million for the Australian Broadcasting Corporation, ranking them number three in earnings among Aussie entertainers. Other characters in this preschool show include the Teddies—three bears of course—, the Rat in the Hat and Kevin the butterfly. B-1 and B-2 are good natured tricksters who sing and joke while offering simple lessons in math, spelling, politeness and proper behavior. One commentator observed that it is a little like Barney meets fruit without the goofy laugh.

After a hiatus of several years, *Bananas in Pyjamas* was planning a return to the small screen in 2002. The new series introduces several new characters including chickens named "Gregory" and "Peck." Perhaps such sly humor is why some adults watch along with their offspring.

For the better part of a century the banana has been part of entertainment, and it is usually comedy. What other fruit could produce the same comedic effect when eaten under water, extracted endlessly from a coat, tied around the waist, piled on the head or worn as a warm and fuzzy costume? Have you ever heard of a "top apple" or "second grape?"

CHAPTER 11

The Musical Banana

"Come Mister tally man, tally me banana,
Daylight come and me wan' go home"
from "The Banana Boat Song" (1956)

A famous American musical personality and his piece of "banana" music first made a big splash in mid-nineteenth century Paris. The Paris of 1845 regarded itself as the headquarters of serious European musical activity, no more so than in music for the piano. The "City of Light," with sixty thousand pianos and one hundred thousand people who could play them, was also the city of pianos. With one-third of the youth and young adults of the city tinkling the ivories, pianos and pianists enjoyed a popularity never before or since equaled.[613]

At the same time the French capital was agog over two American imports. Phineas T. Barnum's Lilliputian wonder General Tom Thumb had appeared at the court of King Louis Philippe and was raking in the cash for the American impresario. Also in the city was American Western painter George Catlin with four hundred paintings of Native American life, eight tons of artifacts and a band of Iowa Indians complete with wigwams, tomahawks and war whoops.[614] Into this environment came a young American musician with several banana associations. He came from New Orleans, Louisiana, a city prominent in the banana trade; he was of Caribbean Creole ancestry; he gained early fame with a musical composition about bananas; and he was musically inspired by time spent in the banana-growing tropics.

The "non-paying" concert debut of fifteen-year old American pianist Louis Moreau Gottschalk grabbed the attention of Paris. On April 2, 1845 the city's musical elite assembled at Salle Pleyel, the most luxurious concert hall in the city. The attendance of Frederic Chopin, perhaps the most celebrated pianist of the time, signaled something special was about to happen. The young Gottschalk performed a demanding program of works by the three leading Paris pianists, Sigismund Thalberg, Franz Liszt and Chopin. After a flawless performance Gottschalk was kissed in the European manner by Chopin who told the young American, "I predict you will become the king of pianists."[615] All agreed the young American was a pianist of exceptional ability.

Louis Moreau Gottschalk's path to Paris started in Louisiana. Moreau (as he was known to his family) was born in New Orleans on May 8, 1829, the first of seven children of Edward and Marie Aimée (Bruslé) Gottschalk. His Jewish father was born in London and educated in Germany before moving to New Orleans in 1820. The elder Gottschalk was a merchant and at times financially strapped. Moreau's mother was Catholic and French in culture and language. The Bruslé family had fled from Saint-Domingue (Haiti) in the mid-1790s after a slave uprising and eventually became part of the Creole population of the multi-cultural New Orleans. It was a cultural environment that provided inspiration for Gottschalk's later piano compositions.

Moreau showed an aptitude for music before he was four, and he took lessons from the organist of the city's St. Louis Cathedral where by 1836 he substituted for his teacher at Mass. By 1842 the family decided that the young prodigy was ready for more intensive training, and he was sent to a private boarding school in Paris run by Monsieur and Madame Dussert. Madame Dussert became a second mother to the thirteen-year-old, and she took him, dressed as Louis XIV, to a masquerade ball where he played the piano.[616]

Moreau's attempt to study at the famed Conservatoire was blocked by Pierre Zimmermann, head of the piano department, who rejected the young American without an audition while commenting that "America is only a land of steam engines." Gottschalk found a place with Charles Halle and later with Camille Stamaty for piano lessons. He studied composition with Pierre Maleden. Along with his musical preparation, the young man was tutored in the ways of a gentlemen. He studied fencing, learned equestrian skills and mastered Italian with the best instructors. After Gottschalk had prepared for three years and attracted attention by playing in fashionable salons throughout Paris, Stamaty decided that the young pianist was ready for his 1845 concert debut at Salle Pleyel.

Despite Gottschalk's success in 1845, it was two and a half years before he performed his first paying concert which took place in the provincial city of Sedan. For the most part the next few years were given to composing works regarded as the exercises of an apprentice. Political upheaval in France in 1847 and 1848 gave the composer's career a forward thrust. The bloody June Days of the 1848 Revolution prompted Gottschalk to follow most of the musical community to safer territory. He spent the remainder of 1848 at Clermont-sur-l'Oise in a relatively safe insane asylum at the invitation of its director.

As a genuine "guest" at the asylum, Gottschalk completed what became known as the Louisiana Trilogy: "Bamboula!," "La Savane," and "Le Bananier "(The Banana Tree). All three pieces had their roots in Gottschalk's New Orleans Creole childhood. Moreau learned Creole songs at home from his Grandmother Bruslé and his African-American nurse Sally. Both women were natives of Saint-Domingue. These were songs that drew upon a French and Afro-Caribbean folk music that was part of Gottschalk's family. His sisters sang them. He learned to play them on the piano, and no amount of time in the French capital could displace them.[617] The name bamboula is the name of a deep-voiced Afro-Caribbean drum, but Gottschalk's "Bamboula" is more than merely reworking a folk song. It is a vital, brilliant virtuoso showpiece. "La Savane" is an artistic vision of a West Indies piece restructured with colorful variations. It closely resembled "Skip to My Lou , My Darling" which probably had the same folk roots.

"Le Bananier, chanson nègre" (The Banana Tree) was based on a well-known march-like song "En avan' Grenadie." " . . . Gottschalk created a little gem, utterly simple at first, but rising eventually to sweeping runs and surging rhythms." The piece begins with what Americans were later to call a "hoochie-koochie" figure in the left hand which sounded like he was setting the stage for the entrance of a belly dancer. Some may detect an Afro-Caribbean influence in the piece, but perhaps it was drawn from the Algerian Casbah as viewed by French troops returning from that North African colony.[618] Later Gottschalk added a fourth Creole piece titled "Le Mancenillier," a reference to a tropical tree that produces small apple-like fruits that are poisonous. These four new, exciting and exotic Creole compositions created a sensation in Paris and Europe.

Gottschalk began programming the Creole pieces in Paris concerts in 1849. "Le Bananier" was performed at a concert on January 11, 1850, and the excited audience demanded an encore. A month later the piece was performed at a concert at Salle d'Erard. The audience cheered for five minutes. The piece was called "one of the most delicious fantasies, and you would have said that a shower of pearls fell in sweetest melody upon the ear."[619] "Bananier" fever raged through Paris. Virtually every pianist was performing it, and one evening the piece was on three programs simultaneously.[620] One Paris critic with typical European rudeness said, "An American composer, good God!" The same critic then praised Gottschalk as a player "of the highest order."[621] Others were more effusive. "It is not the hand of man that touches the keys, it is the wing of a Sylph that caresses them, and causes them to resound with the purest harmony."[622]

With Chopin's demise in 1849, Gottschalk was the virtuoso-apparent. Many could play the piano, but few had the brilliance and drama of performance that marked a virtuoso. "A virtuoso's life was itself a flamboyant work of art. Naturally, the virtuoso did not merely play music but interpreted it."[623] It was a role that Gottschalk was willing and able to play.

In May 1850 Moreau wrote to his father. "All the publishers

declare that no piece ever met with such success as has greeted the Bananier. Already, more than two thousand copies have been sold in Paris alone. It has been pirated in Maycence, Leipzig, Berlin, London, Brussels, Milan, and here, a second edition is being struck off. I played it the past winter at nineteen public concerts, and at sixty grand soirees. [Alexandre] Goria played it eight or nine times, and [Alfred] Jaël plays it everywhere. Benacci [a publisher in Lyons] offered 10,000 francs for the copyright of the Bamboula and Bananier, notwithstanding more than 2,000 copies of the latter had been sold here; my publisher, the Escudiers, answers, 'If you were to offer us 60,000, we should refuse it.'"[624]

By 1850 Gottschalk was a "hot property." Every major piano manufacturer was bidding for his endorsement. In a tour of Switzerland the young American took the country by storm especially with his variations on themes from Rossini's *William Tell.* A Geneva critic thought Gottschalk comparable to Liszt or Thalberg when playing Beethoven's "Sonata in F Minor," but found him a pianist "who will touch you to tears in relating to you on his piano some dreamy legend of his distant country, the 'Bananier,' or the 'Savane,' or in making you behold the African splendor of the 'Bamboula,' that Negro dance."[625] After a concert in Geneva, "the diminutive Gottschalk was waylaid by an ardent lady who picked him up, stashed him in her carriage and galloped off with him. They were not seen again for five weeks." It was but a minor scandal for a virtuoso pianist. After all women were known to battle each other over one of Liszt's abandoned cigar butts. In Lausanne Gottschalk said the audience threw enough flowers on the stage to carpet the theater. He played "Le Bananier" as an encore five times and finally slipped off the stage and "left the lunatics to yell in the desert."[626]

Spain was the next to be conquered, but Gottschalk arrived during a minor diplomatic crisis between Spain and the United States over Cuba. A private concert at the royal palace was arranged for November 17, 1851. Gottschalk's program included "Le Bananier" and "Bamboula," which the pianist had diplomatically dedicated to the queen, in this case a "very tall and stout"

Queen Isabel II who at the time was pregnant with a second baby from a string of lovers. The French-born and impotent King Francisco d'Assisi paid Gottschalk the most polite royal compliment. The King stood in the doorway of the salon and waved as the pianist was forced to retire backwards through a series of four chambers.[627] Gottschalk made himself at home in the palace and, perhaps taking a cue from the Queen, was linked with two romantic affairs. One was with the queen's sister and the other with the future wife of Napoleon III.[628] The pianist was knighted with the Order of Isabel Catolica, presumably for his musical talents.

Months of a grand tour of Spain followed. In gratitude for his reception, Gottschalk composed his largest work to date, the *El Sitio de Zaragoza* (Siege of Saragossa), which commemorated the heroic Spanish defense of the city from Napoleon's retreating armies in 1808. The three hundred page work for ten pianos was written in ten days. The composition which created musical images of the siege was written around a military march and a popular dance. It was premiered on June 13, 1852 at Madrid's Theatro del Principe before an audience of Madrid's aristocracy. In the first part of the program Gottschalk performed "Bananier" and three other works; all were encored. Then came the multi-piano extravaganza. The performance was frequently interrupted by applause. The second part was encored, and when the trumpet and drum sounds of the triumphal march were heard in the third part, the audience rose spontaneously to its feet in an emotional outburst. The Minister of Agriculture in a moment of patriotic fervor, roared, "Long live the Queen." The entire work had to be repeated. The composer was literally carried back to his hotel, and the revelry lasted into the wee hours of the morning.[629] But the ardor, at least that of the royals, cooled, and in November the Queen sent Gottschalk packing for reasons unclear. The pianist returned to Paris briefly before setting out on new conquests.

At age twenty-three Gottschalk set sail for New York in late December 1852. He found that success was harder to achieve in his home land than on foreign soil. Upon arrival in New York he found a city where audiences had "endured a seven-year plague of heavily promoted European divas and thundering pianists" directed by impresarios such as P. T. Barnum. The public was tired of the "pounding virtuosity" of foreign pianists. Only Alfred Jaël from Austria found public favor, and he had performed Gottschalk's "Le Bananier" at his debut. Gottschalk's debut concert on February 11 at Niblo's Saloon also included the work, but it was the first time the American public heard it played as it was intended to be. A "frenzy of enthusiasm" greeted Moreau. One highbrow critic thought that his composition and playing might be "the foundation of a school, at once legitimate, and characteristic. His "Bamboula," "Le Bananier," etc, are truly original specimens of a new and delightful, a purely American, if you please Southern, Creole school, the Gottschalk school, as it may yet be called."[630]

Concerts were seldom solo recitals. Gottschalk almost always got another pianist to perform with him. This assistant artist was usually a worthy performer of lesser prestige thereby enhancing the principal's glory much "in the manner of a five-star general who seems even grandeur when flanked by an escort of three-starrers."[631] A concert by Gottschalk meant mainly music by Gottschalk, and he was often criticized for not playing more Beethoven or Mendelssohn. Gottschalk's explanation was "there are plenty of pianists who can play that music as well or better than I can, but none can play my music half as well as I can."[632] Amen.

Gottschalk made his Boston debut in October 1853. The critics thought the public was "highly gratified" with his unsurpassed technique. Opinion on his compositions was divided. *The Commonwealth* found his "poetic caprices 'La Savane' and 'Le Bananier' had in them much more of caprice than of poetry."[633]

The Boston visit produced an association with the Chickering Piano Company which paid Gottschalk for testimonials and gave him a commission on all pianos he sold as he traveled the country. Considering the piano's popularity, it was a good deal for the pianist. The United States by 1860 reached an annual production of twenty-one thousand pianos of all American makes. That meant that one out of every fifteen hundred persons was buying a new

Gottschalk in 1953 when he returned to the United States after creating a sensation in Europe. (Courtesy New York Public Library, Music Division)

American piano in the course of a year.[634] Gottschalk's Chickering pianos were very durable; they had to be to survive the pianist's often frenetic travel.

Gottschalk's first American tour might have been an artistic success, but it was a financial failure. On top of that his father,

Edward, died in late 1853. Moreau was left with large debts to settle and the need to support his mother and six younger siblings, all of whom had been living in Paris since 1844. With no other immediate possibilities Gottschalk booked passage to Cuba where he spent the next year giving concerts and composing when he could. It was there that he composed the first of several "tear-jerkers."

At mid-century death was a highly fashionable poetic and musical theme, especially if it was the unseasonable death of a child, a sweetheart or mother. Gottschalk composed "The Last Hope," "The Dying Poet," "The Dying Swan," and "Morte!!" (She Is dead). These successful sentimental favorites—the Library of Congress has twenty-eight different editions of "The Last Hope"—tended to produce damp hankies and sometimes hysterics among the women who heard them.[635] When Gottschalk returned from Havana in 1855, these compositions helped ensure many triumphs among the eighty concerts he gave in New York in the 1855-56 season.

Gottschalk embarked on a Caribbean tour in 1857 that became a five year tropical sojourn. Perhaps he was trying to free himself of a long running liaison with an aspiring young actress, Ada Clare, with whom it is alleged that Gottschalk fathered a son. Ada was the acknowledged "Queen of Bohemia" in New York, and according to Walt Whitman, she was "virtuous after the French fashion, namely, [she] has but one lover at a time."[636] Although Gottschalk thought of this period as "lost years," it was a fruitful period in his life as a composer. He later said of the experience, "I have roamed at random under the blue skies of the tropics, indolently permitting myself to be carried away by chance, giving a concert wherever I found a piano, sleeping wherever the night overtook me—on the grass of the savanna, or under the palm-leaf roof of a veguaro [a tobacco-grower] with whom I partook of a tortilla, coffee, and banana which I paid for on leaving in the morning . . . ."[637]

The composer also had reason to complain as he did while in Martinique in 1859. "Save me from these respectable fathers with charming daughters who, in defiance of common sense drum the keyboard from morning to night and make me curse the day when I brought into the world the "Bananier," "Le Banjo," and all the other exotic products that my concerts have brought into vogue in America."[638]

At other times Gottschalk was driven to new heights of creativity as in Havana in 1860 when he organized a "monster concert" at the Grand Tacón Theater. "My orchestra consisted of six hundred and fifty performers, eighty-seven choristers, fifteen solo singers, fifty drums, and eighty trumpets—that is to say nearly nine hundred persons bellowing and blowing to see who could scream the loudest. The violins alone were seventy in number, contrabasses eleven, violoncellos [sic] eleven!"[639] Easily frightened young ladies in the audience of four thousand were advised to bring smelling salts as fainting was expected. The audience heard three new Gottschalk pieces including his symphony *La Nuit des tropiques* [A Night in the Tropics] which was the first composition to employ Afro-Cuban instruments, particularly drums.

By late 1861 Gottschalk was desperate for money; many of his Caribbean performances were done for charity or gratis. He seized a $1500 per month contract offer from Max Strakosch and arranged his return to the United States where the Civil War had begun earlier in the year. As a Southerner, Gottschalk was forced to take an oath of allegiance to the United States before he could enter New York. By doing so he renounced his native Louisiana and his musical roots. His first New York concert was on February 11, 1862 exactly nine years after his 1853 New York debut. On Washington's Birthday in a flag draped Academy of Music he premiered "The Union," which combined elements of his own "Siege of Saragossa" and "Bunker Hill" along with "Yankee Doodle," "Hail Columbia" and the "Star-Spangled Banner" which was not yet the national anthem. The pro-Union Southerner then took his act to Philadelphia, Baltimore and Washington where he was acclaimed even by culturally unsophisticated Congressmen.

Through the entire period of the war Gottschalk kept up a frantic pace and performed more widely than any other artist or entertainer. Along with a company that included the singer Carlotta Patti, Gottschalk swept through the mid-west. Arriving in St. Louis, he saw the arrival of the wounded from the Battle of Shiloh. In Chicago "The Union" excited enthusiasm. Back East he observed the divided loyalties in Baltimore and noted that he could fill the concert hall and make three or four thousand dollars by programming "The Union" and variations on "Dixie's Land." But he thought better of it realizing that with a hall full of partisans on both sides it would come to blows. "It is true in the tumult I might be the first one choked."[640] The entire diplomatic corps attended a concert in Washington. A playful Gottschalk interpolated "The Union" to include the national anthems of every foreign envoy in the house. Each dignitary proudly leapt to his feet and remained at attention as his anthem was played.[641] By the end of 1862 Gottschalk wrote, "I have given eighty-five concerts in four months and a half. I have traveled fifteen thousand miles by train. The sight of a piano sets my hair on end like the victim in the presence of the wheel on which he is about to be tortured."[642]

The ordeal had its high points. In November 1864 Gottschalk performed for President and Mrs. Lincoln. Gottschalk found the President "remarkably ugly, but with an intelligent air, and his eyes have a remarkable expression of goodness and mildness." There were also low points. In Adrian, Michigan, "the people say that they prefer a good Negro show. They are furious at the price for admission—one dollar." When only nine tickets were sold for a concert in Paterson, Gottschalk wrote, "New Jersey is the poorest place to give concerts in the whole world except Central Africa."[643]

Life on the road was an endless procession of dumpy hotels, abominable food, Siberian blizzards and tardy trains. "I live on the railroad," he said. "My house is somewhere between the baggage car and the last car of the train." The war was ever present. "The other day in the car, there being no seat I took refuge in the baggage car, and there I smoked for two hours, seated on the case of my piano, alongside of which, O human frailty! were two other cases also enclosing instruments, now mute, since the principle that made them vibrate, under a skillful touch, like a keyboard, has left them. They were the bodies of two young soldiers killed in one of the recent battles."[644]

By 1865 Gottschalk estimated that he had given eleven hundred recitals and had traveled ninety-five thousand miles. The Far West was the only loyal area Gottschalk had not toured, and so he set sail for San Francisco by way of the Panama Railroad through banana country. In Aspinwall he sought to make a purchase from a young fruit seller who spoke poor English. "I buy some bananas. 'How much?' I say to her. 'Fifty cents,' she answers me. I give her a dollar note, which she returns to me, preferring not to sell to taking paper money."[645] Did she not know this was the composer of "Le Bananier?"

San Francisco sixteen years after the start of the Gold Rush was "a crude and raucous town" culturally split between low brows and high brows. Four out of five males were bachelors and sought entertainment, not culture. Concerts were dismissed as "dearer and less entertaining than other exhibitions," many of which were staged for "gentlemen only." Mark Twain allowed as how Gottschalk got as much out of the piano as was in it which in Twain's words was "tum, tum." Although he liked Gottschalk and found that he could generate "tum, tums" faster and thicker than his landlady's daughter, Twain much preferred the banjo for genuine music.[646]

From San Francisco Gottschalk ventured into the hinterlands. In Virginia City he decided the only gold mine was the hotel where a single room cost $35 a night.[647] For the miners accustomed to being entertained by someone shooting an orange off the head of an assistant, Gottschalk improvised a "battle piece" complete with trumpets and musketry. Packed into stagecoaches along with men with bad breath and muddy boots and corpulent housewives with big trunks, Gottschalk carried "Le Bananier" to Carson City, Dutch Flat, Gold Hills and other places now lost to memory.

Returning to the relative civility of San Francisco, Gottschalk organized a monster concert of fourteen pianos. It was easier to find the pianos than competent pianists. One of the young "artists," the son of the concert's patron, was so "desperately incompetent" that at the performance Gottschalk disconnected the harness of the young man's instrument. Along with the other pianists, the "lad pounded away with gusto." Only after it was too late did he discover that his piano made not a sound.[648]

Gottschalk's stay in San Francisco ended in scandal. Gottschalk was an inveterate womanizer with a long history of conquests. Although he found that California "women are not pretty, and they dress as if the whole stock of the secondhand clothing shops of Paris had been sent to California,"[649] he joined a friend for a nocturnal ride with two young ladies from Mrs. Blake's Oakland Female College that lasted until 2:30 in the morning. The newspaper reported the story two days later without naming the pianist, but rumors spread quickly. On September 18, 1865 in the early hours of the morning a "Mr. John Smith" boarded the U. S. Pacific Mail steamship *Colorado* at the last minute before the ship departed for Panama. The passenger avoided contact with the other passengers. Thus did Gottschalk leave the United States for the last time. Later most newspapers judged Gottschalk guilty of only very poor judgment.[650]

For the next four years Gottschalk traveled and performed in Panama; survived a revolution and earthquake in Peru; avoided a bombardment of Valparaiso and gave a patriotic monster concert in Chile; found a sophisticated musical environment in Uruguay; and survived periodic anarchy in Argentina. On April 21, 1869 Gottschalk began the last journey of his life—this to Rio de Janeiro. He arrived in the active musical world of the Brazilian capital five days before his fortieth birthday. Gottschalk busied himself with performing and composing.

By November Gottschalk was preparing a *concerto monstro* for which he was the composer, fund raiser, conductor, accountant, promoter and soloist. There were six military bands and four orchestras, six hundred and fifty musicians in all. The event left him exhausted. The next night he appeared at a regular concert of the Philharmonic Society. He opened the program with "Morte!!" and then collapsed. He was taken back to his hotel by friends. He was determined to proceed with the second performance of the monster concert the following night, but he again collapsed in pain as he was about to mount the podium.[651] The composer-pianist was constantly attended by a doctor for the next three weeks. At sunrise on December 18, 1869 Gottschalk died of a ruptured appendix—peritonitis—in the village of Tijuca outside Rio. He is buried in Brooklyn's Greenwood Cemetery where no banana trees grow.

"Le Bananier" remained in the repertoire of some pianists for years. The piece was in the library of Bizet. Offenbach played his own arrangement of the piece on the cello. It provided themes for the Polovetsian Dances of Borodin's opera *Prince Igor*. Gottschalk's syncopated music and ragtime style anticipated Scott Joplin, Jelly Roll Morton and other creators of American ragtime.[652] "It is almost possible to draw a straight line from Louis Moreau Gottschalk to Duke Ellington."[653] In 1979 pianist Eugene List assembled forty-two pianists and ten pianos at Carnegie Hall in New York for a "monster concert" to commemorate the 150th anniversary of Gottschalk's birth. They performed "Le Bananier" and other works of the composer. The *New York Times* critic commented that there was a missing ingredient, Louis Moreau Gottschalk himself.[654]

Many decades passed between Gottschalk's "Le Bananier" and the next popular banana song. Until the twentieth century, no songwriters were inspired to write about a fruit that few Americans had seen. When the next banana song made its appearance, it became a wacky world-wide phenomenon. A New York jazz-drummer, Frank Silver noticed a Greek fruit and vegetable peddler who worked the tenement-lined streets of upper Manhattan with a horse drawn wagon. When a customer called from a window for bananas, the immigrant peddler shouted back, "Yes, we have no bananas!" Silver described the funny scene to a fellow musician, Sam Cohn, and together they created their first popular song, "Yes! We Have No Bananas," one of the most successful nonsense songs of the 1920s.

As in many other areas of life in the twenties, Americans were in a mood for songs that were outlandish and absurd with only a remote connection to reality. The "majority of lyrics were sentimental, banal, and either intentionally or innocently infantile."[655] How else to explain "Who Ate Napoleons with Josephine When Bonaparte Was Away?"; "Does the Spearmint Lose Its Flavor on the Bedpost Overnight?"; "Who Takes Care of the Caretaker's Daughter While the Caretaker's Busy Taking Care?"

Silver and Cohn introduced "Yes! We Have No Bananas" in a New York restaurant, but the piece failed to catch fire. The song was rejected by thirteen music publishers before the fourteen publisher took a chance on the novelty with these lyrics:

> There's a fruit store on our street.
> It's run by a Greek.
> And he keeps good things to eat.
> But you should hear him speak.
> When you ask him anything. Never answers "no."
> He just "yes-ses" you to death.
> And he takes your dough he tells you:
> Yes! We have no bananas. We have no bananas today.
> We've string beans and HONions, cabBAHges and scallions
> And all kinds of fruit and say. We have an old fashioned toMAHto,
> Long Island poTAHto, But YES! we have no bananas.
> We have no bananas today.

Some of the song's success came from its discovery by Eddie Cantor. In May 1923 Cantor was starring in *Make It Snappy,* a Shubert brothers plotless musical comedy revue. The show was in Philadelphia and in the last month of its road company run. With very good repeat business Cantor was looking for some new musical material he could use in the show. Visiting the publishing firm of Shapiro, Bernstein and Co. he found "Yes! We Have No Bananas," so new that it was still in manuscript form. Cantor incorporated the song into one of his routines at a Wednesday matinee. The enthusiastic audience response stopped the show cold for over a quarter hour while Cantor sang chorus after chorus. He made the song a permanent part of his act, and it always brought down the house. Cantor's Victor recording of the song became just one of many releases of the piece.[656]

Within a year the sheet music of the song was selling at the rate of twenty-five thousand copies a day. The *Boston Herald* reported that New England alone had purchased one hundred thousand copies—half of them in staid Boston—and fifty thousand copies of the phonograph record.[657] The song enjoyed unprecedented popularity in England, Germany, France and Italy. It was translated into eleven languages including Russian and Chinese. Cantor later stated that when New York Mayor Jimmy Walker came home from Europe a year after the song took off, he insisted that the Europeans thought "Yes! We Have No Bananas" was our national anthem.[658]

As with any success, some questioned the song's pedigree. The *Detroit News* reported from Battle Creek, Michigan:

> Professor Irving Fisher, of the political science department of Yale University, insists that "Yes, We Have No Bananas Today" as diction is admirably stated and that it does not merit the scorn it has received from our mental mandarins and privileged purists. Professor Fischer, a guest at a sanitarium in Battle Creek, Mich., was asked by a *Detroit News* correspondent whether the sentence was correct English. The professor stroked his academic beard, ruminated in a scholarly manner for a while, then smilingly said, "Yes, it is correct—on a certain hypothesis." The reporter, strangely enough, was eager for more information. "Well," said Professor Fischer, "it is misleading at all times; yet, however, it is technically correct in answer to the question: 'Do you have no bananas?'"[659]

Dr. Charles Dwight Reid, an American physician at St. James Hospital in Anking, China had a Chinese translation made which Mr. George Allan England, of Boston, took to his friendly laundryman for a critique. The laundryman delivered an indignant verdict. "It say, today, yes, banana, have not got. It say, bean, onion, have got. Everything, cabbage, little onions, have got. Everything, fluit [sic], have got. Say so. Red melon, potato from Kwow Wow, have got. It say, but, today, yes, banana, have not got. Who

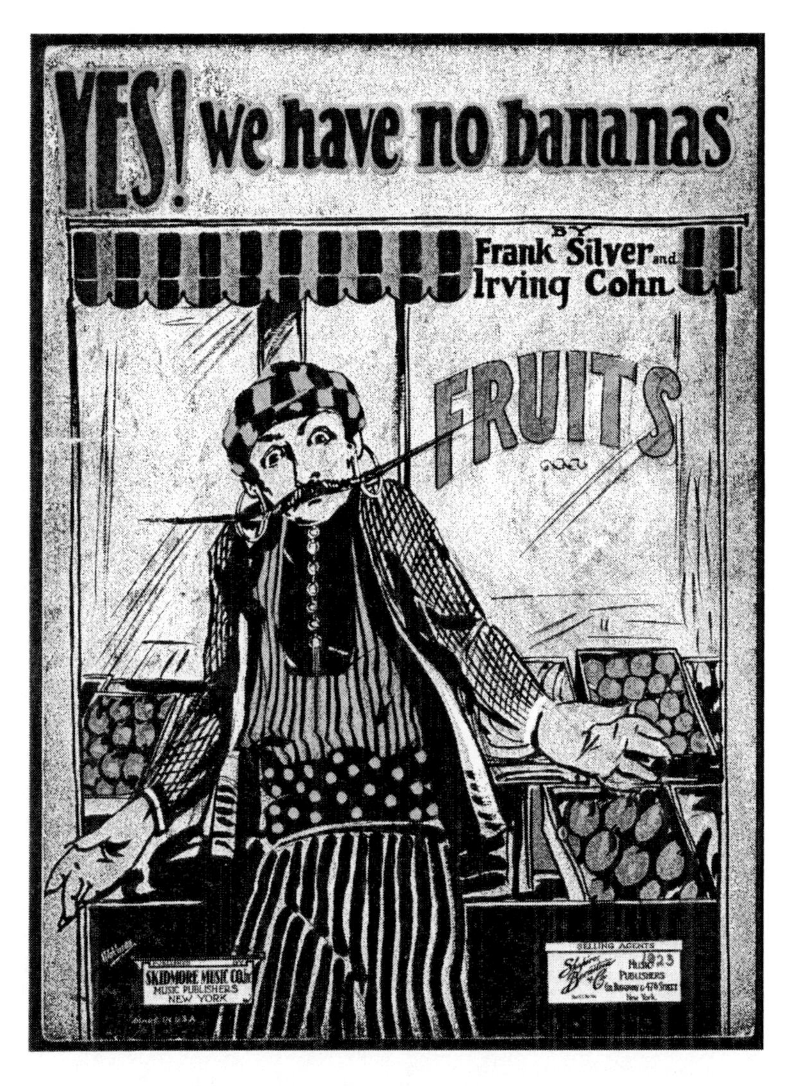

The 1923 sheet music for "Yes! We Have No Bananas. At one time after its appearance, it was reported to be selling twenty-five thousand copies a day. (Courtesy New York Public Library, Music Division)

write damn fool Chinese? Where you get 'em? Clazy [sic] mans! Yes, sir. No good!"[660]

In *Music Box Review of 1923* the tune was parodied in a performance where it was done in a grand-operatic style of the sextet from *Lucia di Lammermoor*.[661] The roots of the song may not have been operatic, but they were deep. Paul Whiteman wrote in 1926 that the art of composing comes in knowing how to steal and how to adapt. "Pretty much everybody knows now that Handel's *Messiah* furnished the main theme of the well-known 'Yes, We Have No Bananas,'" said Whiteman. The very first bars of the chorus can be found in Handel's work, and if the words that go with these bars from the *Messiah* were to be sung in the banana song, you would find yourself starting the chorus with "Hallelujah" Bananas. Whiteman also wrote that "perhaps it is not such general knowledge that most of the banana song which wasn't taken from the *Messiah* came from Balfe's 'I Dreamt That I Dwelt in Marble Halls.'"[662] Another source attributes part of the song to "The Vacant Chair"—or "We Shall Meet But We Shall Miss Him,"—a 1861 melody that referred to the death of Lt. John Wilson Grout of the 15th Massachusetts Volunteers Infantry.[663] A further probe of the piece's genesis reveals the last line from "My Bonnie Lies Over the Ocean," Cole Porter's "An Old Fashioned Garden," and the line "I was seeing Nellie home" from "Aunt Dinah's Quilting Party."[664] Remarkably, there is not a fruit song in the bunch.

With one foot on a banana peel, the *Boston Herald* suggested a conspiracy was afoot. "There are evidences on every side that this song is an insidious plot against our reason," said the paper.[665] Conspiracy or no, everyone jumped on the fruit wagon. Frank Silver, a co-conspirator, organized a ten-piece "Banana Band" and packed dance halls and theaters throughout the United States. With a pictured field of bananas as a backdrop, each musician was seated before a large bunch of real bananas. Said Mr. Silver, "I know when I'm lucky."[666]

The two novice songwriters must have known that they had reached the big time when Will Rogers claimed "it is the greatest document that has been penned in the entire History of American Literature" and if the writers had written the National Anthem, "it wouldn't take three wars to learn the words." The great humorist wrote that the subject was real genius because, "when you get down and write of Cabbages, Potatoes and Tomatoes, you just about hit on a Universal subject." He predicted an epidemic of corned beef, liver and bacon, soup and hash songs to flood the market.

Rogers observed that "Mother has been done to death in Songs and not enough consideration shown her in real life. We thought when we sang about her we had paid her all the respect there was. I tell you conditions were Just Ripe for a good fruit song." Rogers went on to point out the merits of each verse of the banana song. "you know, it don't take much to rank a man away up if he is just lucky in coining the right words. Now take for instance Horace Greeley, I think it was, or was it W. G. McAdoo, who said, 'Go West, Young Man.' Now that took no original thought at the time it was uttered. There was no other place for a man to go, still it has lived. Now you mean to tell me that a commonplace remark like that has the real backbone of this one: 'Our Grapefruit, I'll bet you, Is not Going to wet you, we drain them out every day.' Now which do you think of, that or 'Go West, Young Man?'"

The final Rogers heaping of praise—if that be the right word—was his observation that "I would rather have been the Author of that Banana Masterpiece than the Author of the Constitution of the United States. No one has offered any amendments to it. It's the only thing ever written in America that we haven't changed, most of them for the worst."[667]

"Yes! We Have No Bananas" has enjoyed a sustained presence in film. The song was interpolated in the Al Jolson movie musical, *Mammy* in 1930. It was sung by the Pied Pipers in *Luxury Liner* in 1948, and in 1954 the voice of Eddie Cantor was heard singing the song on the soundtrack of *The Eddie Cantor Story*.[668] In a 1996 film audiences hear the song sung with gusto by a group of explorers/mappers of the Royal Geographic Society around a campfire in the North African desert. In film time it was somewhere in the 1930s and Hungarian Count Almasy, played by Ralph Fiennes, knew the words to "every song" of the period. The film is writer-director Anthony Minghella's Academy Award winner for best picture, *The English Patient*.

After "Yes! We have no Bananas" the next big banana song was

"Chiquita Banana" which the United Fruit Company made into an advertising jingle in 1944.

The original song, written in 1938 by Len Mackenzie, Garth Montgomery and William Wirges, experienced several incarnations. The best known version is the marketing jingle that educated consumers on how to ripen bananas and warned not to put them in the refrigerator. In another version Chiquita tells the story of how she came from a little equatorial island on a big banana boat to help good neighbor policies. She says she could sing about moonlight on the tropical equator but sings of bananas and refrigerators instead.

A folk song sung to the tune of "Chiquita Banana" has nothing to do with refrigerators. Its lyrics say:

> I'm Chiquita Banana and I'm here to say
> How to get rid of your teacher the easy way
> Just eat a banana, throw the peel on the floor
> And watch your teacher slip out the door.[669]

The first Chiquita who told the world to keep bananas out of the refrigerator was Patti Clayton who performed the tune with the Ray Block Orchestra. Clayton as a singer had a "good, swingy voice with perfect clarity of diction." The song as a radio advertisement soon became No. 1 on what *Time* magazine called the "jingle-jangle hit parade." The first recording remained in circulation for eight months. The United Fruit Company began to print and make gratis distribution of the sheet music.[670]

Elsa Miranda (no relation to Carmen), a twenty-four year old stenographer who could not read a note of music and had never taken voice lessons, was "discovered" when she sang at an office party. She recorded the first Spanish version of "Chiquita Banana," and went on to sing the song in English in rhumba, tango, samba and calypso versions.[671] The song became a jukebox, disc-jockey, and big-band hit and was interpreted in one fashion or another by Fred Allen, Charlie McCarthy, the King Sisters, Xavier Cugat, Ellery Queen and Arthur Fiedler.[672]

"Chiquita Banana" was often pressed into service for special causes. In 1945 the lyrics were changed to aid famine relief in Europe. Americans were asked to eat more fresh vegetables and fruits—which of course includes bananas—so that more fats, wheat and canned goods could be sent overseas. In 1949 and 1965 the jingle's ending was changed from "No, no, no,no!" to "Don't waste a drop!" to enlist public cooperation during New England water shortages.[673]

Like "Yes! We Have No Bananas," "Chiquita Banana" made a modern film cameo appearance. Woody Allen, who seems to have a "thing" with bananas, used the song in *Everybody Says I Love You*, a 1996 film that he wrote and directed. This comedy-musical starred Allen, Alan Alda, Drew Barrymore, Goldie Hawn, Julia Roberts and Tim Roth. In a Halloween scene three trick-or-treaters, a little girl in a banana costume and two boys in south-of-the-border garb, deliver an enthusiastic rendition of "Chiquita Banana." Banana songs, it seems, just don't go away.

In 1997 Chiquita Brands International decided that the lyrics of the banana song needed a little updating, and the company sponsored a contest to write new lyrics that stressed the banana's health benefits. The winning lyrics were written by Joyce Appelquist, a fifth grade teacher from San Meteo, California.

In 1999 Chiquita again updated the song to emphasize the nutritional aspects of the fruit using these lyrics:

> I'm Chiquita Banana and I've come to say
> I offer good nutrition in a simple way.
> When you eat a Chiquita you've done your part
> To give every single day a healthy start.
> Underneath the crescent yellow
> You'll find vitamins and great taste
> With no fat, you just can't beat 'em
> You'll feel better when you eat 'em.
> They're a gift from Mother Nature and a natural addition to
> your table
> For wholesome, healthy, pure bananas look for Chiquita's
> label![674]

The "Chiquita Banana" song has an enduring quality; so too

does a banana song popularized by Harry Belafonte. It may not be the way that he wants to be remembered, but Belafonte will always be known as the artist who brought the world "The Banana Boat Song" with its famous opening line, "Day-O, Dayyy-oh. Daylight come and me wan' go home." The world first heard Belafonte's performance of "The Banana Boat Song" on the RCA Victor album *Calypso* released in the fall of 1956. The album became a best seller by January 1957 and stayed on the best seller list for over a year and a half. It was the first RCA LP by a single artist to sell over a million copies. Today the song remains the single biggest number in his repertoire, and Belafonte often invites concert audiences to join him in closing the show with the calypso song. "'That song is a way of life,' he says with little-boy enthusiasm. 'It's a song about my father, my mother, my uncles, the men and women who toil in the banana fields, the cane fields of Jamaica. It's a classic work song.'"[675] Belafonte experienced Jamaica early in his life.

The "King of Calypso"—an appellation he doesn't like—was born Harold George Belafonte on March 1, 1927 in Harlem in New York City. Both parents were from the West Indies, his mother from Jamaica and his father from Martinique. In the Harlem community the young Belafonte saw a "do-rag" coiffed Duke Ellington shopping for fish; Joe Louis striding along 125th Street; and Langston Hughes patronizing a local bar. But Harry's mother decided that the poverty and the dangers of the city were too much for her children, and in 1935 she sent them to live in Jamaica. There he got to see another poverty—that of the plantation owned by British absentee landlords. "'I saw black men down at the docks who were singing work songs like 'Day-O' as they loaded and unloaded boats,'" says Belafonte. "'When I sing 'The Banana Boat Song,' some people glimpse it as a fanciful little tale that charms and delights the listener. But for the singer, it talks about a human condition that was very real to me, very painful and extremely oppressive.'"[676]

Before the "Day-O" that Belafonte popularized there was the traditional "Banana Boat Loader's Song," a "digging" song as were all Afro-West Indian songs to accompany work. It was a rhythmic chant of the men who loaded bananas on ships in the cool of the night. The song's roots were in calypso, a musical style that originated in Trinidad. Slaves in the eighteenth century Trinidad sugar fields were forbidden to talk by their Spanish overseers. Singing was permitted because the rhythms seemed to add to productivity. Thus slaves could converse, spread gossip, voice grievances, and plot possible revolt in song. Calypso songs were sung first in French patois which was carried by emigrants from Haiti. They were not sung in English until the beginning of the twentieth century. The tradition of story-singing was well established by the time slavery was abolished in Trinidad in 1838. Typical calypso lyrics were witty, full of satire and double-entendre. The sexual double entendre of a song such as "Big Bamboo" is thinly veiled. Calypso lyrics were frequently improvised on the spot to comment on a current issue such as the latest scandal, corruption, sporting event or personal experience. The lyrics were usually repetition of verse and chorus led by a leader known as a "Calypsonian."[677]

Calypso was not accompanied by instruments until it moved out of the work fields. Drums were the first accompaniment but were banned in the mid-nineteenth century. Bamboo poles of varying lengths and thicknesses were struck on the ground until these "tamboo bands" were also banned in the 1920's because the poles were used in crime and gang wars. At the time of World War II percussion instruments were cleverly fashioned from abandoned petroleum drums, and from that development the well-known "steel bands" emerged.[678]

African rhythms were common in all West Indian music. The African tribes represented in Spanish and French colonies differed from the tribes in English colonies. The calypso songs that Belafonte heard in Jamaica derived almost completely from the Akan tribes of the Gold Coast area of Africa.[679] But two decades passed before Belafonte drew upon these calypso roots to record "The Banana Boat Song."

In 1940 Belafonte returned to Harlem. His mother told him he should not be like his Uncle Lenny, then in the Harlem numbers racket, but he should aspire to be like actor-singer Paul Robeson. But with dyslexia, it was a big task, and he dropped out of school after ninth grade. He joined the Navy at age seventeen and inadvertently received a radicalized education; he discovered W. E. B. DuBois when someone gave him a copy of *Color and Democracy*. It made him an angry young man and sparked a continuing interest in improving the life of African-Americans.[680]

After the Navy Belafonte returned to Harlem and worked as a janitor's assistant. Someone gave him tickets to a performance by the American Negro Theater. "'The curtains opened," remembered Belafonte, "'and out on the stage walked these black people, telling a story, and it had rhythm; it had a purpose; it had a mission. I was absolutely overwhelmed, and I thought, 'Jesus Christ, what a hell of a place.'" He was hooked. He used the G. I. Bill to enroll in drama classes at the New School for Social Research. Among his classmates were Marlon Brando, Walter Matthau, Rod Steiger, Tony Curtis and Bea Arthur.[681]

His acting debut was at the A.N.T. in an Irish play, Sean O'Casey's *Juno and the Paycock*, with a black cast. Belafonte found the struggles of the Irish and the West Indians to be much alike. Paul Robeson saw a performance and complimented the young actor; with a mutual admiration the two men became friends.

Hearing Belafonte's singing voice, a friend in the drama workshop coaxed him into appearing at an amateur night at the Royal Roost in New York where he quickly landed a regular gig. Starting with jazz he also moved into performing folk and calypso songs. Success took him to the Village Vanguard and a recording contract with RCA Victor.

In 1953 Belafonte appeared on Broadway in *John Murray Anderson's Almanac* for which he won a Tony Award. He later completed an artistic "hat-trick" with both an Emmy for his 1960 television special "Tonight With Belafonte" and a Grammy in 1985 for organizing the album and video *We Are the World* which supported African famine relief.

Also in 1953, Belafonte made his film debut in *Bright Road* opposite Dorothy Dandridge with whom he also appeared the following year in *Carmen Jones*, an all black version of Bizet's opera. The latter film is now considered a classic. But it was Belafonte's third record album that brought him lasting banana fame.

The *Calypso* album which included "The Banana Boat Song" was cut in the Fall of 1955, but the song was first heard by American audiences on October 2, 1955 on the NBC Colgate Comedy Hour in a twenty minute "Holiday in Trinidad" segment. Calypso singer-composer Lord Burgess—his real name is Irving Burgie—of West Indian descent composed the song along with Bill Attaway, writer of the TV show's script, and Belafonte. But the song did not catch on until the album was released in 1956.[682]

Another version of the traditional song was written by Erik Darling, Bob Carey and Alan Arkin in 1956 and aroused popular interest as a single recorded by The Tarriers. Belafonte's single of "Banana Boat" followed, and by February 1957 both singles were in *Billboard's* "Top Twenty:" The Tarriers at number six and Belafonte at number seven. By April, when Elvis Presley's "All Shook Up" was number one for eight weeks, Belafonte's "Banana Boat" was number three and went on to outsell the Tarrier's version in the United States and the world. For a while some observers speculated that calypso might displace rock and roll.

A Belafonte craze swept the country. Pete Seeger noted, "'Day-O'—it's something that just makes you feel like taking a lungful of air and singing. A lot of people who never thought of themselves as singers could sing it.'"[683] Some calypso purists complained that Belafonte was not an authentic calypsonian, a charge that Belafonte freely admitted. He said he interpreted the music for the American audience in such a way as to project a positive image of the Caribbean region. The critics hardly mattered. *Look* magazine wrote that Belafonte had become the first Negro matinee idol in our entertainment history. He went on to record twenty-eight more albums.

For many years Belafonte was absent from the Hollywood scene because he could not find a script that was not racially offensive. He turned down *To Sir, With Love* and *Lilies of the Field*, the 1963 film that won the Academy Award for his friend Sidney Poitier.

Belafonte had no major film role between 1974, when he played in *Uptown Saturday Night*, and 1995, when he starred with John Travolta in *White Man's Burden*, a social satire based on racial role reversal. In 1996 he won critical acclaim in Robert Altman's *Kansas City* in the role of Seldom Seen, a gangster owner of a nightclub. His long film hiatus was put to good use.

Harry Belafonte has always been a campaigner for civil rights. In 1957 when a white landlord would not rent him an apartment in Manhattan's Upper West Side, he bought the apartment building. It wasn't the last time he would use his artist's income to make a point. Belafonte first met Dr. Martin Luther King, Jr. in 1956 during the Montgomery bus boycott, and it was Belafonte who bailed King out of the Montgomery jail. The singer's financial resources funded many civil rights causes. He provided seed money for the Student Nonviolent Coordinating Committee. Belafonte was a trusted advisor until King's assassination, but the singer's fight against social injustice at home and abroad continued unabated.[684] In addition to efforts for African famine relief, Belafonte has also served as a special advisor to the Peace Corps and as an ambassador for UNICEF in defending children around the globe.

Now in his 70s and with a slightly raspier voice Belafonte still delivers the "Banana Boat Song" with the same charisma and dazzling energy as four decades earlier. And thanks to Tim Burton's movie *Bettlejuice* in which a malicious ghost forces a staid dinner party to dance to a calypso beat and sing "Day-O" with abandon, a whole new generation discovered "The Banana Boat Song."

While Belafonte's "Banana Boat Song" is at one end of the banana distribution system—Jamaicans loading bananas aboard ship, Harry Chapin's "30,000 Pounds of Bananas" was near the other end—a young truck driver delivering a load of ill-fated fruit to Scranton, Pennsylvania. Both the Belafonte and Chapin songs tell a story in the folk song tradition, and like Belafonte, Chapin was known to give his time and financial resources to good causes.

Chapin's story-song, written in the late 1960s and always a show-stopping concert piece, tells the tale of a young tractor-trailer driver on only his second run delivering a day's worth of bananas to Scranton. From almost any direction it's all downhill into the coal-scarred city where children play on slag piles. Thinking about the woman who awaited him at journey's end, the young driver did not notice the warning to shift to low gear for the curving two mile run from the top of the hill. As the city's lights twinkled below, he anticipated the evening's delights, but when his foot touched the brake, it gave no response. As the thirty thousand pounds of bananas picked up speed, he could only utter the name of the only force that could save him. He entered the steepest grade at ninety miles an hour, passed a fortunate bus and prayed it was all a dream. Before he came to a dead stop he sideswiped nineteen cars, snapped thirteen poles, hit two houses, smashed eight trees and "Blue-Crossed" seven people. The unfortunate man never saw the thirty thousand pounds of bananas smeared over the length of four football fields.[685] Did it really happen?

Harry Chapin's biographer says Chapin was an unsung hero. "He was an American Jacques Brel, a modern-day Woody Guthrie, the Pete Rose of popular music. He was a spokesman of the people, the poet laureate to cab drivers, housewives and common folks alike."[686] Like the troubadours of old, Chapin used simple melodic and rhythmic forms to tell the everyday sometimes tragic tales of ordinary people. Chapin loved losers. In 1972 in his first hit, "Taxi," a man bemoans his unfulfilled dreams. "W.O.L.D." is a tale of a disc jockey's wasted life, and "Sniper" is the story of the real man who killed people at the University of Texas. In perhaps his best known song, the top-selling 1974 hit "Cat's in the Cradle," a man reflects on missed paternal opportunities with his son. The composer-singer was always loyal to the little guy.

"30,000 Pounds of Bananas" was released in 1974 on Chapin's fourth album, *Verities and Balderdash*, a title which may or may not be a clue to the song's basis in fact. The piece was written originally as a story-poem and later a story-song, a process Chapin used for many of his songs. According to his biographer, after 1967 "he began to write with complete constructs, creating environments he'd never been in that weren't necessarily factual in detail, but truthful in emotion."[687] Chapin himself said, " . . . I feel , as in all my perverse songs ("Sniper," "30,000 Pounds of Bananas"), they were struggles of people toward a life force, even though they may take strange ways." And the inspiration for "30,000 Pounds?" Chapin explained. "It was triggered by my reaction to Vietnam body counts that accentuated statistics rather than humanity. . . . Ironically enough, nobody ever sat down and added up all those figures because we were claiming enough casualties to have wiped out the population of Vietnam twice."[688] Now we know. Or do we?

Life sometimes imitates art. Chapin's banana truck drive may never have existed, but an unlucky thirty-six year old driver from Tacoma, Washington did. Hauling a load of forty-four thousand pounds of bananas from California to Tacoma, the driver took a wrong turn on his way to the Interstate, and lost control of his rig on a curvy mountainous grade. The driver and the bananas—perhaps slightly bruised—survived.[689]

Harry Chapin was not as lucky. He was killed in an automobile accident on New York's Long Island Expressway not far from home on July 16, 1981. He was a young thirty-eight, and ironically, like the young truck driver in "30,000 Pounds of Bananas," he was killed by a tractor-trailer. It struck his Volkswagen Rabbit from behind on the day he was to begin a three month tour starting with a benefit concert.

Some things about Harry Chapin were unmistakable; one of them was his social conscience. In 1987 Harry Belafonte hosted a benefit to fight world hunger on what would have been Chapin's 45th birthday. The evening was filled with anecdotes of the man. Bruce Springsteen admitted that he sometimes tried to hide if he saw the activist-songwriter coming. Otherwise he was sure to be drawn into a long conversation about the state of the world. Springsteen also recalled that Chapin always said, "I play one night for me and one night for the other guy."[690]

He received the Rock Music Award for Public Service in 1976 and 1977 and the B'nai B'rith Humanitarian Award in 1977. He was a founding member of the World Hunger Year and a member of President Carter's Commission on World Hunger. In 1987 Congress voted Chapin a posthumous award of the Congressional Gold Medal for his efforts in raising money and awareness to combat world hunger. Thirty thousand pounds of bananas would help.

From Gottschalk's "Le Bananier," the banana tree, through Belafonte's Jamaican workers of the "Banana Boat Song" and the immigrant fruit peddler in "Yes! We Have No Bananas" to Chapin's truck driver in "30,000 Pounds of Bananas," ordinary people in the banana business are represented. All except the consumer for whom there are plenty of banana songs to go around. Try "I Like Bananas Because They Have No Bones." Or you might like "Apples, Peaches, Bananas and Pears," "Bananas and Cream," "Banana Pudd'n." "Banana Split" or "Bananas By the Bunch." For romantics there is "Banana Baby," "Banana Love," "Banana Split for My Baby" and "Loving You Has Made Me Bananas." For dancers there is "Banana Boat Limbo" and the "Banana Cha Cha." The politically inclined will enjoy "Banana Republic," "Ode to the Banana King" and "All the Nations Like Bananas." For those with weird relatives we have "My Brother Thinks He's A Banana" and "My Wife Left Town with a Banana." And remember "Please Don't Squeeza Da Banana." Wow! Yes! We do have banana songs—bunches of them.

CHAPTER 12

Celebrating the Banana

"To me the banana is the international symbol of smiles, good health and a positive attitude in this fast-paced going-bananas world."
Ken Bannister, 1998

Food is big, literally! World records for big food have been set regularly in recent years. 1997 was a banner year for egg dishes. In November the French created the world's largest quiche Lorraine in Paris. It had an eight-foot radius. In the same month the Thais prepared the world's largest omelet using twenty-one thousand eggs. Thai hens must have been exhausted. America is the land of "big." So it was fitting that a Boy Scout troop in Minnesota put together a three-ton popcorn ball, and in August 1998 San Francisco weighed in with a world record two ton Caesar salad.[691] But the banana was ahead of them all.

For more than thirty years beginning in 1963 the good citizens of the twin cities of Fulton, Kentucky and South Fulton, Tennessee annually prepared a humongous banana pudding. A one ton banana pudding which fed ten thousand visitors was the centerpiece of Fulton's annual late summer fun-and-games Banana Festival. When Fulton's two thousand pound concoction was challenged by a three thousand pound banana pudding in Canada, Fultonians doubled their efforts and produced a four thousand pound monster pudding.

Kay Martin, Executive Director of the Chamber of Commerce, had a major responsibility of directing the Banana Festival from

1984 until the last festival in 1992. She vividly remembers her first introduction to the town and the festival. "When I moved to Fulton, which was in 1973, we came into Fulton in the middle of the banana festival. And I moved to Fulton from Columbus, Georgia. It was quite shocking to come into town, a small town like this in extreme Western Kentucky, and see stalks of bananas hanging on the parking meters. That was my real introduction. I thought, wow, these people have lost it." But Martin quickly caught the spirit of the festival and the pudding. "To me one of the most interesting things about the festival was making the pudding."[692]

Martin's enthusiasm was shared by Elaine Forrester who was involved in the Festival from its start. A real hometown girl, Forrester's father was a barber and her mother worked at the Brown Shoe Company in town. Forrester has served as the mayor of Fulton and owns a beauty salon and a floral shop, conveniently located next door to the funeral home. Forrester too has fond memories of the pudding. "People were lined up to do it every year. People looked forward to doing it. It was fun; it was fun."[693]

The idea for a banana pudding came from W. P. "Dub" Burnette, the head of a local dairy, during the planning of the first festival. He proposed the idea to Jo Westpheling, another planner, who thought that he would need at least a "tub full of the stuff." "I told her 'tub full' nothing, I'm thinking about a ton," said Burnette.

Burnette took his wife's grandmother's recipe for boiled custard and expanded it by trial and error, kept careful records of weights and measures and ended up with a four pound pudding. He multiplied those ingredients by five hundred to create a one ton banana pudding.[694]

For those who wish to serve ten thousand guests with a one ton banana pudding, you will need three thousand bananas, not too ripe nor too green, peeled and sliced, two hundred fifty pounds of vanilla wafers, and nine hundred fifty pounds of boiled custard, prepared ahead of time and cooled. The bananas for Fulton's banana pudding were usually donated by companies such as Dole and Chiquita. Nabisco contributed the vanilla wafers. When the local dairy said they could no longer supply the boiled custard, the Regency Pudding Company of Louisville stepped in with banana crème pudding packed in five-gallon buckets.

A one ton pudding requires more than your everyday bowl. "Dub" Burnette contacted the Pittsburgh Glass Company and was told that they could produce a suitable glass container for $20,000. A Memphis company supplied a less pricey plexiglass container for the one ton banana creation. When the Festival produced a two ton pudding, the Modern Welding Company of Owensboro, Kentucky built and donated a six hundred fifty pound metal tank that was six feet long, three feet deep and three and one-half feet wide, considerably larger than the average bathtub. Kay Martin remembers it well. "They brought it down to us, and it just took an act of Congress to get it unloaded, the two ton of pudding in it, and then haul it around by a tractor."[695]

Martin remembers constructing the pudding. "That was a lot of fun to get all of the ingredients together and get it coordinated and get the local people in there with the rubber gloves and little hair net things on and slice three thousand bananas." The bananas were sliced by pressing them through a device strung with wire. "You just had people everywhere," recalls Martin. "We would get the vanilla wafers, and you would empty the boxes into the trash bags. Somebody would take a bucket of bananas and layer them. After you did one layer of bananas for a one ton pudding, you had to go and change gloves because I mean it is really messy. It was a lot of fun."[696] Elaine Forrester also has vivid memories of the assembly process. "There were some people who wanted the peels, but I don't know what they did with them. I know some people wanted the peels to put on their flowers. We had to put it together in the refrigerator truck and freeze to death. After you stir and bend over that for a while, you get warmed up. It was fun."[697]

The monster banana pudding made its grand appearance in the festival parade Saturday morning. A forklift was used to move the pudding onto the trailer that would be pulled by a tractor in the parade. The operation was sometimes perilous. "Dub" Burnette remembers one time coming down a small incline at the dairy when two thousand pounds of pudding almost spilled onto Fourth Street.

At the end of the parade the pudding was served to the hungry multitude. Elaine Forrester was a regular server. "We kept the plastic sealed on top of it and cut out holes to dip it out with," she recalls. "I had blisters on my hands and bruises on my legs from being up against the metal. It was cold, but it also bruised your legs because you leaned up against it so long. We tried to serve everybody before we let them come back with their buckets. People knew how good it was, and they would take it to the nursing homes and to the people who had to work." The health department man was always close by, and when the pudding reached a certain temperature, it had to go.[698]

Some people never heard of a banana pudding much less a one ton specimen. Most members of the U. S. Marine Band appearing at the festival one year had no knowledge of the banana delight, but they were anxious to return once they had experienced it. Then there was Miss America 1963, Donna Axum, who as an honored festival guest was the first to be served. She wondered aloud, "Are you really supposed to eat this stuff?" "Bud" Burnette told her, "do what you like, but we made it to eat." Taking him at his word, the beauty queen tasted and squealed with delight, "Oh, it's good!"[699]

The one ton banana pudding was but one aspect of the International Banana Festival that enjoyed a thirty year run in Fulton—South Fulton. If you visit the twin cities today you will find no evidence of any banana presence beyond what you would find in any small town in the United States. The two towns with a combined population of about six thousand straddle the Kentucky—Tennessee state line.

The Fulton Theater on Main Street is now closed, but most of the business is located on nearby Lake Street anyway. Like many downtowns there is evidence that the big action is now out at the malls on the highway. Attractive green and yellow banners offer a "Welcome to Fulton" and signal a town with an upbeat attitude. The Marigold Furniture store is closed, but the nearby Rentavision store offers appliances and furniture. Just down the street Sonny and Vada Puckett's Cafe Capone offers attractive sidewalk tables under an awning in the colors of the Italian flag. The City National Bank building with its four massive columns lends an air of stability that dates back to 1906. Kay's Style N' Go, Cissy's Gifts and the Evans Rexall Drug Store offer their services. Cross Lake Street and an unused railroad track bracketed by park and parking, and it is just a few steps to Tennessee where the Browder's Seed and Grain Co., originally the Browder Milling Co.—maker of "Queen's Choice" flour—crowds the border.

There is still an agricultural presence in the area. It is predominantly soy beans, but there is also some tobacco, some rice, and some cotton in the delta area along the Mississippi River about twenty miles west of Fulton. There are some automotive related businesses and some clothing manufacturers in the area. The Ferry-Morse Seed Company is the largest business in town. But there are no bananas grown around Fulton. So why was the banana celebrated by a festival in Western Kentucky and Tennessee? The simple answer is the railroad.

Fulton lies within a territory known as the Jackson Purchase because Andrew Jackson, acting for the state of Kentucky, bought the land from the Chickasaw Indians in 1811. Fulton's first settlers were of English descent and arrived by way of the Carolinas. In the 1850s a post office called Ponotoc was established in the area. For a time the place was referred to as "The End of the Line" by the U. S. government and the mail was addressed in that way. Eventually, the residents relinquished the Indian name Ponotoc to the town of Ponotoc, Mississippi and renamed their community Fulton in honor of Robert Fulton the famed American developer of the steamboat.[700] But it was the railroad not the steamboat that put Fulton on the map.

Construction of the New Orleans and Ohio Railroad, incorporated in 1852 by the citizens of Paducah, Kentucky, was built south to the Kentucky—Tennessee state line in late 1860. A station named Fulton was erected at a point that crossed an extension of the Mississippi Central Railroad. Eventually, both rail companies became part of the Illinois Central System which became the principal north-south railroad linking New Orleans with Chicago through the Mississippi River valley. By the 1930s the Illinois Central had more than five hundred connections with over one hundred fifty railroads.[701] When one looks at the Illinois Central System on a map, one sees a pattern of lines that resembles an hourglass. Fulton occupies the narrow point through which almost all of the rail traffic must pass.

The Illinois Central was the first land-grant railroad and promoted colonization of its territory on a large scale. The road was a pioneer in promoting agriculture in its territory. As early as 1868, the line was carrying entire trains of refrigerator cars of southern Illinois fruit and vegetables to markets around the nation. Around that time the Illinois Central started the "Thunderbolt Express," the first all strawberry train in America. It ran from southern Illinois during the strawberry season. In 1908 when the boll weevil threatened cotton fields in Louisiana and Mississippi, for an entire month the railroad sent agricultural experts on an educational train over its lines in Mississippi. Over ten thousand people heard lectures on this train known as the "Boll Weevil and Diversified Farming Special." In 1927 experts from the agricultural college of Illinois visited thirty-five thousand farm families and businessmen on the "Soy Bean Special." The education effort worked. Between 1927 and 1934 there was an average increased production of over one million bushels of soy beans in Illinois.[702] With a strawberry and soy bean friendly railroad, could bananas be far behind?

Allegedly, the connection between bananas, the Illinois Central Railroad and Fulton began with a Civil War romance. When the Union Army occupied New Orleans, a handsome, twenty-two year old Union colonel named James T. Tucker set the hearts of Crescent City maidens aflutter. The ambitious "Jimmy," as he was known to his friends, was smitten by the dark-eyed Leontine Piseros, one of the most charming Creole belles of the city. When the war ended and Jimmy returned to his former position with the Illinois Central Railroad, he could not forget the vivacious Leontine and managed to get himself appointed the railroad's general agent in New Orleans. Jimmy was aggressive in both the business of the railroad and courtship of Leontine. Fending off other suitors, he won the hand of the lovely Leontine, and they married on Mardi Gras eve in 1870.

As we know, bananas were beginning to appear on New Orleans wharves in the 1870s, and the astute Jimmy Tucker began to think of how the yellow fruit would produce rail revenue when carried up the line to Americans who did not yet know what they were missing. Alas, poor Jimmy was unable to realize the banana's potential. He was struck down by a fatal illness in 1874 but not before he had alerted his brother Joe to the banana's possibilities. Joe became the railroad's traffic manager in 1877, and by 1880 the company handled a grand total of twenty-two carloads of bananas out of New Orleans. In 1881 the number was up to 331 carloads and for the first time exceeded rail shipments of coconuts. By 1900 it was approximately eight thousand carloads, and in 1920 it was 28,478 carloads out of New Orleans. The all-time record of hauling 52,757 carloads was reached in 1947. The Illinois Central was the nation's greatest banana carrier and much of the yellow fruit went through Fulton.[703]

In the early twentieth century refrigerator railroad cars such as this carried thousands of loads of bananas through Fulton each year. (Courtesy Library of Congress, LC-USZ62-105216)

Fulton laid claim to the title "Banana Capital of the World," and it was said that seventy percent of America's bananas passed through the city. The latter claim appears to be a misinterpretation of the statistics, but there is no doubt that huge quantities of the fruit passed through Fulton en route to the nation's heartland. The title "Banana Crossroads of the United States" is more appropriate because the Illinois Central owned the two main lines that crossed at Fulton, and the city became the principal Mississippi Valley stop for the perishable freight inspection service. Resident and traveling messengers regularly took the temperature of the delicate fruit in shipment and then adjusted the heating or icing as conditions required. The R. H. Wade Ice Company and then the Fulton Ice Company iced the banana cars after their five hundred mile northbound trip from New Orleans or Mobile.

From this banana heritage Fulton's International Banana Festival emerged in 1963. Jo Westpheling may have started the whole thing in 1958 when as the sponsor of a local high school singer and ukelele player appearing on the Arthur Godfrey Show she proclaimed Fulton the banana capital of the world. The young ukelele player made additional appearances on the Godfrey show and each time Westpheling plugged Fulton's bananas. Letters began to roll in from perplexed people all over the country wondering how a small town in Kentucky could be the world's banana capital. Nathan Wade, known as the furniture king of Western Kentucky, crafted the idea of a banana festival to promote the area. If Bells, Tennessee could hold a Okra Festival and Paris, Kentucky an annual Fish Fry, Fulton could benefit from a banana festival. In the summer of 1963 Wade, Westpheling and others journeyed to New Orleans to meet with an association of banana dealers. The Fultonians asked for $15,000 from the banana men and promised to raise another $15,000 back home. The banana men said yes; the locals quickly subscribed their share; and everyone in Fulton was seeing yellow—or maybe it was tourist dollar green.[704]

With the work of hundreds of volunteers from the twin-cities, financial support from state tourism grants and an attendance of ten to twelve thousand visitors, the festival took off. From the beginning it was an international festival of friendship between the people of the banana growing Latin American nations and the people of Fulton. An early festival adopted the theme of PROJECT

UNITE-US which had a Cold War era goal to "fight communism with bananas." Actually, blankets became the first weapon of choice. When it was learned that indigent mothers in some Central American countries left hospitals with new-born babies wrapped in newspapers, a "Brigade for Baby Blankets" was organized to collect hundreds of baby blankets for a hospital in Quito, Ecuador. Miss America of 1963 loaded the first bundle of blankets to send them on their way to Quito via truck and banana boat.[705]

Each year the festival joined forces with Operation Amigo to bring twenty-five to fifty Latin American students to Fulton for two weeks. In 1972 two hundred Honduran students and their escorts attended. Families in Fulton and South Fulton hosted the teens who were exposed to all aspects of life in the United States from schools to rodeos. Kay Martin remembers the Guatemalan Army Marimba Band which was a big hit at several festivals. "It was a lot of fun. They brought all their xylophones and instruments. We would haul them around on the back of a flatbed truck, and they would play at the park and downtown."[706] Elaine Forrester recalls of the marimba band that "some of them got to come more than one year and got to be real good friends. I sometimes wonder what happened to them because I know they had some unrest in their country."[707]

The festival was sometimes the scene of domestic and international diplomacy. At the 1972 festival Kentucky Governor Wendell Ford challenged Tennessee Lt. Governor John Wilder in the Banana Shoot that required contestants to split a half-banana in two using a bow and arrow. Ford lost, but the performance of Mayor Jose Fernandez of San Pedro Sula, Honduras was diplomatically not reported.[708]

The highlight of the four day Banana Festival was always the Saturday Parade in which the one ton banana pudding occupied the same position of honor than Santa Claus occupies in the Macy's Thanksgiving Parade, the end. The pudding and barbecue was served to thousands in the city park at the conclusion of the parade. But what happens if it rains on your parade, and you have prepared a pudding and have barbecue for thousands in the Chamber of Commerce office? And what if the downpour becomes a flood and turns Lake Street into what its name implies? "Well the first thing," according to Kay Martin, "is ten thousand people calling and wanting to know are you having the parade." They didn't. With water coming in the front door, Festival officials got on the phone to spread the word that barbecue was available by the pound. Just row up to the front door for your barbecue and your buckets of free pudding. Says Martin, "We turned into a meat operation."[709]

To fill the schedule of a four-day festival, you need a lot of events, and Fulton had no shortage in that department. There were banana contests of every kind including a Banana Bake-Off, a Banana split eating contest and a Banana Bonnet contest—for both sexes. Of course there had to be a contest to find an International Banana Princess. Elaine Forrester, the Festival President in 1978-9, was a participant in the first Banana Princess contest. A Banana Bowl football game was started in 1987 and pitted the Fulton High School Bulldogs against the South Fulton Red Devils.

Arts, crafts and theater added another dimension to the Banana Festival. Works of both American and Latin American artists were exhibited with some regularity. Mayan Indians demonstrated weaving, and natives of Ecuador demonstrated their prowess with the machete, the tool much used in harvesting bananas. Theater ranged from the sublime to the ridiculous. A junior company of the Joffrey Ballet and musicians Peter Nero and Lionel Hampton graced Festival stages. Touring theater companies performed and gospel music groups sang. Then again, there was the "Chicago Knockers All Girl Mud Wrestling Team" which commanded the highest admission price at the 1987 festival.[710]

Civic spirit reached a fever pitch at Festival time. In 1979 when the flags of the banana producing countries needed replacing, the homemakers clubs made nine new ones to fly on Lake Street. Volunteers ran the Banana Academic Bowl, the road race, the book sale and the multitude of other events. Businesses, such as the Dairy Queen seized the moment and offered banana splits at "peeled back prices." And there were fireworks including one episode when the program began with the detonation of four large mortar devices improperly aimed. The thunderous booms send the shocked crowd fleeing from the field. As Kay Martin remembers it, "We were trying to find the guys who ordered those because if that was the beginning, we could only imagine that the end would knock the stadium down. It got better as the night went on."[711]

Some events may have been fun but of questionable taste. Kay Martin's son recalls with restrained pride that as a teenager he participated in the tobacco-spitting contest without getting sick. Then there was the one time "Cow in the Square" event concocted by a creative local disc-jockey recently arrived from Michigan. Martin recalled with amusement how the creator presented the idea. "We are going to get a cow and put up temporary plastic fencing around the parking lot at the end of Lake Street. We are going to draw squares, put the cow in there, and sell chances on where the cow will do its thing. Whoever has that square gets the pot." A big crowd was on hand for the event. Perhaps it was stage fright, but whatever it was, the cow would not accommodate. Nevertheless, the crowd had a lot of fun buoyed perhaps by visits to "The Keg," a next door establishment which dispenses what its name implies. "Everybody was down there," said Martin, "waiting for the cow who wasn't going along with the scheme at all. So we had to go and get the local vet." The encounter with the vet moved the cow to declare a winner.[712]

The Illinois Central Railroad no longer carries bananas, the Fulton ice plant burned down more than a decade ago, and the last of thirty International Banana Festivals was held in 1992. The end wasn't the cow's fault; it was a case of a dwindling number of volunteers and the difficulty of raising money to cover rising costs. Says Mayor Forrester, "I wish we still had it because it was a fun time. It does require a lot of work, and it was year round work." The Banana Festival was replaced by "Weekend at Ponotoc," a scaled down festival but not banana related. The one-ton banana pudding was made for a few more years but is no more. During its run the Banana Festival generated community pride and cooperation, promoted international friendship and understanding, and put the Twin-Cities on the tourist map. Oh, and yes it served a lot of banana pudding—at least sixty-four thousand pounds of the stuff. That's a lot of bananas.

If banana fame has left Fulton, it has moved to Altadena, California, the home of the International Banana Club and Museum. The Top Banana of the bunch is Ken Bannister (a.k.a. Bananister), an affable ambassador of humor for, in his words, "the bright yellow, curvaceous, elongated fruit from the herbaceous plant." Bannister is a professional photographer whose permanent smile suggests he must have said "cheese" once too often. To know him is to laugh.

Good humor is what the Banana Club is all about. "Its purpose is to keep people smiling, their spirits up," Bannister says with a grin. "It's used as a vehicle to do the same thing all over the world." There are no rules, regulations or regular meetings for the club that has about nine thousand members. The club didn't start; it evolved. In 1972 Bannister's secretary's husband, a stevedore who unloaded banana boats, gave him a roll of Chiquita stickers. Bannister started handing out the stickers to people at photo trade shows. Bannister recalls the reaction, "Everyone smiled at me when I handed them their banana sticker as I went around the country, and folks started sending me things to do with bananas."[713]

For more than thirty years people have joined the bunch and have chosen special titles. There are C.H.Ds., City Hand Directors; R.R.Bs., Real Ripe Bananas; and a TB for Bannister, the self-dubbed Top Banana. Bannister's family has bought into the fun. His wife Chris is the TBW., Top Banana's Wife, and Bannister's three daughters Lori, Julie and Cheri appropriately carry the titles TBD. I, II and III respectively.

With members in twenty-eight countries, the club is truly international. There are lots of members in Germany—you know those Germans and their bananas—, Sweden, Norway, Finland, Australia and Japan. "You know," says the TB, "I have a Russian. That's his title, Russian Banana, R.B. He has got all of Russia. What a franchise!"

The ever "up-beat" Ken Bannister, the "Top Banana" of the International Banana Club in the Banana Museum. (Courtesy Elliott Marks Photography)

The Banana Club counts presidents of nations among its members. The former president of Ecuador, Abdala Bucaram, joined through the club's web site (www.banana-club.com). If humor is the hallmark of the club, President Bucaram was the perfect member. Elected the President of his country in 1996 after an outlandish road show which featured Bucaram as a singer, dancer and comedian, he referred to himself as "El Loco." Several weeks after he joined the Banana Club, the Congress of Ecuador removed him from office for "mental incapacity."[714]

The honorary title P.B., Presidential Banana, was conferred on President Ronald Reagan. Several years after granting Reagan the honorary membership, Bannister met the President when he was photographing a fund-raiser in Los Angeles. He asked the President, "Do you remember me sending your Banana Club kit to you?" A perplexed Reagan replied, "Weeellll, well, banana man. No." The TB send the P.B. another kit. Most recent Presidents have been honorary club members, but somehow or other, President Clinton managed to escape the honor. Said Bannister, "I'm still analyzing his smile. I analyze the genuine smile. I don't get into politics. So, I won't go there."[715]

As Bannister spread his stickers and the banana items rolled in—they now number more than eighteen thousand—,the TB began to award BMs, banana merits, which members can use to earn two degrees including a Ph.B., Doctor of Bananistry. Everything must be in good taste. "Nothing rude, crude or lascivious," says Bannister." Among the banana items collected is a highly prized rock-hard petrified banana sent in by a photographer from Kentucky. "That was worth one hundred banana merits," says Bannister. "Bang, she got a degree like that. She puts M.B. Master of Bananistry, after her name." Sitting on a big yellow banana couch in the museum, the TB remembers that a furniture manufacturer once asked him, "Ken, how many BMs do I get for a couch?" "I said automatic Ph.B degree. Let's have it." Ray Harrington of Minnesota, who among other things sent in an original recording of "Yes, We Have No Bananas," has the most banana merits of anybody.[716]

Where do you keep eighteen thousand banana items? "Back in 1976," Bannister says, "I had to rent a place because I was looking a little strange in my office and at my home with all these banana items. I would like to preserve everything that everyone has given, sent, contributed to the museum from all over the world. They can come see it here." The Banana Club Museum is a storefront in Altadena and is open to members and to others by special arrangement. There have been a few unexpected intruders such as the curious fire truck from the fire station across the street that smashed in the front door several years ago to see what all the fun was about.

If, as the psychologists say, the color yellow stimulates the intellect and creativity, to enter the Banana Club Museum is to become an instant raving genius. "It's pretty much overwhelming, isn't it?" asks Bannister. "The yellow and the banana paraphernalia." A field of goldenrod in full bloom doesn't have as much yellow. A visitor who finds a spot to sit on the yellow banana couch becomes just a part of the bunch in the "soft" section of the museum where several hundred stuffed, plush smiling bananas of every size along with banana slippers and the banana muppet from Jim Henson's company hold forth. Sitting behind a desk covered with banana lamps, banana phones—including a cellular banana—, a banana stapler, banana pens and paper and the banana yellow pages, Bannister alternately demonstrates the banana nose and the banana ears and observes that "I have enough items here and in the drawers and cabinets in my home probably to fill the Getty Museum downtown. Tastefully, and you know displayed properly. You can see where things are a little bit bunched up in here."[717]

The banana pictures alone could fill most museums. As expected, there are pictures of Carmen Miranda and Josephine Baker. Then there is Tim Conway dressed as a banana, Marilyn Monroe with more curves than a banana, John Wayne with a banana in his holster, and the Statue of Liberty holding a banana aloft. "You can see if things were framed like old Harvey Korman here," says Bannister, "if things like that—original ads, and works of art, photographs—were framed properly, we would fill up the Getty Museum with no effort at all." If the TB has his way, the Banana Club Museum will expand into more luxurious quarters as part of casino in Las Vegas, a concept he hopes to sell to Donald Trump.[718] Can you get three bananas on a slot machine?

Meanwhile the papier-mâché bananas, wooden bananas and glass bananas—clear and colored—accumulate in the art section where a highly valued piece of stained-glass art catches the eye. It got the maker an automatic degree. The boudoir banana from Italy is a plastic banana attractively decorated with lace and sequins. Another sequined number was called the Michael Jackson banana by its donor. A banana needlepoint was created by a "Little Old Lady in Pasadena"—her Banana Club title.

The food section has banana popcorn, banana marshmallow pies, banana gum candies from Japan, banana cookies and banana O's, a takeoff on Cheerios. Banana type cereals and candies are in great abundance. There is banana-cinnamon tea that can be brewed in the banana teapot. The banana drinks include Crème de banana and a Japanese breakfast banana drink that resembles "V-8—at least that is what the TB thinks the Japanese lady said."

There is no shortage of banana personal products including banana pipe tobacco, suntan oil, toothpaste, soap and, for kids who want their hair to stick up like crayons, banana hair putty. The "Macho Banana Body Spray" has been around for over eighteen years and probably smells like it. The banana perfume is not recommended by the TB.

You can use banana wax on your car and spray "banana slice repellent" on your golf balls before you whack them with your banana putter with the banana shaped head. Sam Snead posed with one, but Johnny Carson on the "Tonight Show" never did swing the one Bannister sent to him.

The Banana Club Museum must be one of the few museums in the world to have a refrigerated section which protects such items as a little chocolate banana car and banana popsicles. It is also where Bannister keeps the banana candle that can be burned at both ends.

Among the collection of banana cartoons is one that causes Bannister to chuckle as he reads the caption on the cartoon that shows two cowboys, one of whom has a bunch of bananas over his face. The cowboy says, "I distinctly said bandanas." "Wow, I can't believe how long I have been doing this," says Bannister as he recalls that the cartoon arrived in 1978. "People will send things, anything that they see in the newspaper or in a magazine. They are all welcome. I don't tell people that I get duplicates because that's not polite. That's ok; they're all worth a banana merit or more. I'm pretty generous with the banana merits because they make people feel good."[719]

Banana Club degrees are awarded each summer at the annual B. C. picnic and games. Recipients of the Doctor of Bananistry degree receive a medal with their name inscribed. "This is big stuff especially when you wear this on an airplane," says the TB. The picnic and games held each year in Southern California is a guaranteed three to four hours of laughter for the entire family. The bananas are supplied by the banana companies or supermarket chains. There is the obligatory banana eating contest—the record is sixteen six ounce bananas in two minutes—and the banana pie eating contest. In the "Draw and Peel" each contestant pulls a banana in six-shooter fashion from a pocket and peels it. The first naked banana wins. The "Banana Hole-in-One Putt" uses the famous Banana Club putter. The hole is about six to eight inches across, and because the ball is placed only two feet from the cup, nobody misses. Even the least athletic members can do well in the banana put—like the shot put but a lot less weight. Says Bannister, "Senior citizens can come and get a first prize ribbon like everybody else. We give everybody a ribbon as first prize so nobody's feelings get hurt." Kodak often supplies prizes for the games. "I couldn't go to Fuji," says Bannister, "because Kodak was yellow. Anything that's yellow." Even picnic hot dogs are yellow. Actually, they are vegetarian dogs because they are bananas with a little whipped cream on top. Overall, it's a tasteful family gathering.[720]

In 1998 the Banana Club picnic and games was part of the California State Fair in Sacramento. At the invitation of the State, Bannister appeared as "Banana Man," and he conducted the games four times a day over an eighteen day period. Four hundred items from the Banana Club Museum were displayed in the children's building at the fair.

For Bannister bananas have become a vocation. Whether it's doing a half-hour show at a very cramped Banana Club Museum for twenty-three third-graders; giving a lecture on the funny fruit at a service club; or appearing with Jay Leno on the "Tonight Show," the TB is sure to be wearing something yellow with the smiling club logo. The yellow banana jumpsuit with peels—a club member earned five hundred merits and an instant degree for making it—suits supermarket openings. For meeting Presidents a low-key, tasteful cloisonné Banana Club pin fits the bill. For everything in between there is a home closet full of golden apparel, and if that is not enough, there is the collection of four or five dozen banana T-shirts, shorts and other garments in the museum.

A Top Banana has to have a bananamobile. In this case it's an older creamy-yellow BMW, complete with "TB" license plates, which Bannister's wife now drives. Someone from Northern California once said they had an actual banana race car. Says Bannister, "I said how much do you want for it, and the guy said, 'a hundred and fifty thousand dollars.' I said I'll be sending you a check right away." Bannister did good-naturedly offer to accept the car as a donation to the museum.[721]

Jay Leno described Bannister as a "banana evangelist." Few would disagree. Teachers in dozens of schools award students Banana Club stickers for achievement instead of the usual gold star. Two of Bannister's daughters who are teachers gave him the idea. The TB uses Banana Club gift memberships to promote achievement in schools and to improve spirits in programs and homes for the elderly and less fortunate. Someone once called him the morale officer of the world. "There isn't anybody that I can't get to smile," he says.[722]

Ever the proselytizer, Bannister supports the efforts of others enamored by the banana. An independent bunch puts on a banana ski trip to Aspen, Colorado each year. Yellow garbed skiers schuss down the slopes. Every year in Hilton Head, South Carolina several thousand people compete in the banana open tennis tournament organized by Dennis Malich. "He's the eastern TB," says Bannister. "He wanted to split the country with me. I said you can have as much as you want."[723]

It was inevitable that the two major celebrants of the banana, Fulton's International Banana Festival and the TB of the International Banana Club, should hook up. Bannister was invited to the festival twice, in 1980 and 1987. The TB proudly displays the Top Banana plaque presented to him by the Banana Festival. "They rolled out the yellow carpet and put me in the parade with my yellow suit," says Bannister.[724] Fulton's mayor, Elaine Forrester, recalls Bannister's visit. "He had a good time," she chuckles. "He is quite a character."[725] There is no disputing that.

The banana man is on a mission. "When I go somewhere, I've

got this smile on my face," says Bannister while demonstrating of what he speaks. "Hopefully, it will bring one to their face, because you know life is too short. It's a little stressful these days and everybody is going nine hundred miles an hour. So this is a great reminder to keep smiling and stay in good health."[726] What better reminder than the fruit with the permanent smile—the banana.

Acknowledgments

Although the research and writing for this work was accomplished in the past five years, the idea for this book first saw the light of day about fifteen years ago when I was still teaching. The encouragement of many colleagues, friends and acquaintances kept the light focused on the ultimate goal for all that time.

The four research libraries of the New York Public Library are a priceless treasure house of resources. The librarians and staff of the New York Public Library for the Performing Arts, the Center for the Humanities, the Schomburg Center for Research in Black Culture and the Science, Industry and Business Library were consistently helpful and efficient. The same is true of the staff of the Library of Congress, especially the Prints and Photographs Division. In New Jersey the staffs and resources of the Somerset County Library in Bridgewater and the Bernards Township Library in Basking Ridge were invaluable.

Many good friends read and criticized portions of the manuscript at various stages of its preparation. I am especially indebted to Katherine Venditti who read multiple versions of the work with a keen eye for both form and content. Thanks also to Jack Beaver, James Farrell and Mary Ann Farrell whose suggestions along the way improved the final result. The comments and suggestions of numerous other friends helped direct my efforts.

Perhaps strangely, encouragement was found in rejection letters from publishing house editors who found the idea to be "wonderfully quirky and fun," "intriguing" and "fresh and funny." One editor's response that the firm was not considering "fiction or poetry" still remains a mystery.

As the politicians would say, "I accept full responsibility" for the final text, and any errors are mine alone.

Dennis N. Fox
Basking Ridge, New Jersey
August, 2002

Cover: The cover character, "Banana Andy" created by Marianne Tremont; cover photography and author photo courtesy of MHP photo

Permission to reprint images and excerpts is gratefully acknowledged as follows:

From *Spock on Spock: A Memoir of Growing Up With the Century* by Benjamin Spock, M.D. and Mary Morgan. Published by Pantheon Books. Copyright © 1985, 1989 by Benjamin N. Spock and Mary Morgan. Reprinted with permission of Lescher & Lescher, Ltd. All rights reserved.

From *Gravity's Rainbow* by Thomas Pynchon. Published by Viking. Copyright © 1973. Reprinted with permission of Penguin Putnam Inc.

For sheet music cover image and lyrics. ***Yes, We Have No Bananas*** by Frank Silver and Irving Cohn. Copyright © 1923 Skidmore Music Company, Inc., New York. Copyright renewed. International Copyright Secured. All Rights Reserved. Used by Permission.

For lyrics, *The Chiquita Banana Song*. Lyrics copyright © 1999 Chiquita Brands International, Inc. Music © 1964 Shawnee Press Inc. Used by permission.

NOTES

Chapter 1 Notes: The Ubiquitous Banana

[1] U.S. Census Bureau, *Statistical Abstract of the United States.* 2000.

[2] Roger Yepson, *Apples*. (New York: W. W. Norton and Co., 1996), 14.

[3] "Your Cousin, the Banana," *Discover* (Jan. 2001), 62.

[4] "America's No. 1 Second Banana Once 'Sold All Over Washington,'" *Washington Times* (October 29, 1986).

[5] *Detroit News*, (January 5, 1996)

[6] *Fruit Dispatch*, (March 1921), 445.

[7] "Banana-peel Lawsuits," *The Christian Science Monitor* (Feb. 16, 2002), 2.

[8] Patti Hagan, "Behold the Banana, Sub Rosa Breakfast of Champions." *Wall Street Journal* (July 14, 1994), A-8.

[9] Alex Abella. *The Total Banana* (New York: Harcourt Brace Jovanovich, 1979), 30.

[10] Trevor Hellems. Interview with the author. Cranford, N. J., May 31, 1997.

[11] *New York Times* (February 22, 1942), III, 3.

[12] "Bananas," *Sports Illustrated* 82 (Feb. 8, 1995), 18.

[13] Jay Coyle, "Yes, We Have Bananas: Things That Go Bump In the Night When You're Aboard." *Yachting*. 179 (June 1996), 16.

[14] Carole Potter, *Knock on Wood, An Encyclopedia of Talismen, Charms, Superstitions and Symbols.* (New York: Beaufort Books, 1983)

[15] Martha Beckwith, *Hawaiian Mythology*. (Honolulu: University of Hawaii Press, 1970), 17-18.

[16] W. Mieder, *Dictionary of American Proverbs.* (1992), 36.

[17] "Banana Splits," *Restaurant Hospitality* 80 (March 1996), 84.

[18] *Guinness Book of World Records, 1997.* Mark C. Young, ed. (Guinness Publishing, Ltd.), 1996.

[19] "A Banana Republic: Chefs Go Ape Over Fruit's Desserts," *Nation's Restaurant News* 26 (June 13, 1994), 37.

[20] "Banana-Rama," *Restaurant Business* (April 1, 2000), 81.

[21] Judith Martin, *Miss Manners' Guide to Excruciatingly Correct Behavior.* (New York: Warner Books, 1983), 142, 150.

[22] Letitia Baldridge, *Amy Vanderbilt Complete Book of Etiquette.* (Garden City, N. Y.: Doubleday, 1978) 29, 426.

[23] Steven E. Landsburg, "The Great Banana Revolution," at www.slates.com, June 27, 2002.

[24] "Mrs. Brown's Diamond Ring," South Carolina Writers' Project, Project 1655, September 25, 1939. *WPA Life Histories Collection*. (Washington: Library of Congress).

[25] "Brooklyn Streets," October 17, 1938. *WPA Life Histories Collection* . (Washington: Library of Congress).

[26] Irenee Du Pont, *America's Castles.* Arts and Entertainment Television Network, January 26, 1997.

[27] *New York Times*. (March 2, 1997) 1.

[28] *Courier News.* Bridgewater, N. J. (August 17, 1997), A-1

[29] *Hartford Courant.* (May 19, 1997), B-5.

[30] Clive Gammon, "Banana Republic's Survival Chic is Wining Bunches of Trendy Buyers," *Sports Illustrated.* 63 (August 19, 1985), 88; "How to Find Your Market," *Changing Times*. 39 (September 1985), 32.

[31] *New York Times.* (August 23, 1996), Sec. 9, 1; *WWD.* (July 10, 1997), 23.

[32] *WWD*, (Sept. 4, 1998), 9.

[33] Mark Landler, "Now, Worse Than Ever! Cynicism in Advertising," *New York Times*. (August 17, 1997) Sec. 4, 1.

[34] *New York Times*. (April 17, 1994) F-25; Mary Lou Zimmermacher, Hewlett-Packard Public Relations Office, January 27, 1997.

[35] "Bananas: Development in the West," *The Economist*. 327 (April 17, 1993), A-27.

[36] Duane Rhodes, "Chrome Banana," *Common Ground.* (Spring 1997), 47.

[37] *People Weekly*. 25 (June 16, 1986), 85.

[38] *The History of the Maine Bear*, (University of Maine at Orono, n.d.).

[39] *Hartford Courant.* (September 20, 1990).

[40] *Tampa Tribune*. (February 10, 1991).

[41] "Standing Up for the Banana," *Harper's Magazine*. 275 (December 1987) 20.

[42] Thomas P. McCann, *An American Company: The Tragedy of United Fruit.* (New York: Crown Publishers, 1976), 84.

[43] *Newark (N.J.) Star Ledger*, (Dec. 29, 1997), 7; Marianne Lavelle, "Food Abuse: Basis for Suits," *National Law Journal*, (May 5, 1997), A-1.

[44] *Time*, (April 27, 1998), 85; *People Weekly*, (April 27, 1998), 13.

[45] "Not the Elephant's Child," *The Economist*. 324 (August 8, 1992), 74-75.

Chapter 2 Notes: The Banana Family Tree

[46] Dr. Herbert J. Spinden, "The Early History of the Banana," *Fruit Dispatch* (Aug. 1926), 212.

[47] Book review of Peter Hatch, *The Fruits and Fruit Trees of Monticello: Thomas Jefferson and the Origins of American Horticulture*, (University Press of Virginia) in *Preservation* (May-June 1998), 103.

[48] Doug Richardson, interview with the author, Feb. 25, 1997.

[49] Richardson interview.

[50] Richardson interview.

[51] Richardson interview.

[52] N. W. Simmonds, *Bananas* (London: Longmans, 1959), 105.

[53] Richardson interview.

[54] Although the horse originally evolved in the New World, it was unknown in the Western hemisphere when Columbus arrived in 1492.

[55] Claire Shaver Haughton, *Green Immigrants: The Plants That Transformed America* (New York: Harcourt Brace & Jovanovich, 1978), 32.

[56] Viola J. Herman, and Carolyn Margolis, *Seeds of Change* (Washington: Smithsonian Institution Press, 1991), 141, 166.

[57] *Ethnobotany: Evolution of a Discipline*, Edited by Richard Evans Schultes and Siri von Reis, (Portland, Oregon: Dioscorides Press, 1995), 9.

[58] Jared Diamond, *Guns, Germs, and Steel: The Fates of Human Societies*, (New York: W. W. Norton and Company, 1997), 119-122.

[59] Charles Bixler Heiser, *Seed to Civilization: The Story of Food*, (Cambridge, Mass.: Harvard University Press, 1990), 154.

[60] Bruce D. Smith, *The Emergence of Agriculture*, (New York: Scientific American Library, 1995), 142.

[61] M. D. Kajale, "Mesolithic Exploitation of Wild Plants in Sri Lanka: Archaeobotanical study at the Cave Site of Bali-Lena," in *Foraging and Farming: The Evolution of Plant Exploitation*, Edited by David R. Harris and Gordon Hillman, (London: Unwin Hyman, 1989), 269-277.

[62] Smith, 36-41.

[63] Les Groube, "The Taming of the Rain Forest: a Model for Late Pleistocene Forest Exploitation in New Guinea," in *Foraging and Farming: The Evolution of Plant Exploitation*, 299-301.

[64] *Origins of Agriculture: An International Perspective*, Edited by C. Wesley

Cowan and Patty Jo Watson, (Washington: Smithsonian Institution Press, 1992), 4.

[65] R. H. Stover and N. W. Simmonds, *Bananas*, 3rd. ed., (New York: John Wiley and Sons, 1987), 263-7.

[66] Stover, 83.

[67] Stover, 272-3.

[68] S. G. Harrison, *The Oxford Book of Food Plants*, (London: Oxford University Press, 1969)

[69] Philip Keep Reynolds, *Earliest Evidence of Banana Culture*, (Baltimore: American Oriental Society: 1951), 6, 14.

[70] Simmonds, 54.

[71] Reynolds, *Earliest Evidence*, 9-10.

[72] Philip Keep Reynolds, *The Story of the Banana*, (Washington: Government Printing Office, 1923), 24.

[73] Frank Browning, *Apples*, (New York: North Point Press, 1998), 64-65.

[74] P. K. Reynolds and C. Y. Fang, "The Banana in Chinese Literature," *Harvard Journal of Asiatic Studies*, 5(June 1940), 165-167.

[75] H. Edward Hill and John Evans, "Crops of the Pacific: New Evidence from the Chemical Analysis of Organic Residue in Pottery," in *Foraging and Farming: The Evolution of Plant Exploitation*, 423.

[76] E. Cecil Curwen and Gudmund Hatt, *Plough and Pasture: The Early History of Farming*, (New York: Henry Schuman, 1953), 222-3, 225, 227.

[77] Diamond, 148.

[78] Diamond, 400; Harrison, 196.

[79] Robert L. Hall, "Savoring Africa in the New World," in *Seeds of Change*, Edited by Herman J. Viola and Carolyn Margolis, (Washington: Smithsonian Institution Press, 1991), 166.

[80] Food and Agriculture Organization, *Roots, Tubers, Plantains and Bananas in Human Nutrition*, (Rome: Food and Agriculture Organization, 1990), 71; Curwen, 247, 249, 251-2, 261, 264.

[81] Jack R. Harlan, "Indigenous African Agriculture," in *Origins of Agriculture: An International Perspective*, Edited by C. Wesley Cowan and Patty Jo Watson, (Washington: Smithsonian Institution Press, 1992), 68.

[82] *Oregonian*, (March 17, 1995).

[83] Hall, 163-6; Reynolds, *Earliest Evidence*, 19.

[84] Reynolds, *Earliest Evidence*, 26-27; Simmonds, 57.

[85] Reynolds, *Earliest Evidence*, 21.

[86] J. Smartt and N. W. Simmonds, *Evolution of Plant Crops*, 2nd ed., (London: Longman Scientific and Technical Books, 1995), 372.

[87] Diamond, 132.

[88] Peter N. Davies, *Fyffes and the Banana: Musa Sapientum - A Centenary History 1888-1988*, (Atlantic Highlands, N. J.: Athlone Press, 1990), 9.

[89] *Summary and Recommendations of First Scientific Meeting of the Banana Improvement Project,* (Washington: Agricultural Research and Extension Group, The World Bank Group, 1996), available on internet.

[90] Richardson interview.

[91] Richardson interview.

[92] William O. Lessard, *The Complete Book of Bananas*, (n.p., 1992), 1.

[93] Stover, 167-168.

[94] Richardson interview.

[95] Seaside Banana Gardens Consumer Information Sheet; Lessard, 41, 83.

[96] Seaside Banana Gardens Consumer Information Sheet.

[97] Seaside Banana Gardens Price List; Lessard, 47.

[98] Lessard, 35; Seaside Banana Gardens Consumer Information Sheet.

[99] Lessard, 65, 67.

[100] Lessard, 39, 55, 61.

[101] Smartt, 373; Robert A. Read, "The Banana Industry: Oligopoly and Barriers to Entry" in Mark Casson, *Multinationals and World Trade: Vertical Integration and the Division of Labour in World Industries*, (London: Allen and Unwin, 1986), 322.

[102] Lessard, 57.

[103] Thomas P. McCann, *An American Company: The Tragedy of United Fruit*, (New York: Crown Publishers, 1976), 72.

[104] Larry Luxner, "Super Banana," *Americas* (English Edition), 45 (Nov. - Dec., 1993), 4.

[105] Richardson interview.

[106] *Wall Street Journal*, (April 10, 1995), A1, A13; *Discover*, 15 (July 1994), 22.

[107] Michelle Hibler, "Breeding a Better Banana," *International Development Research Centre Reports* (January 1998)

[108] *Christian Science Monitor*, (Nov. 14, 1996), 6.

[109] Andy Coghlan, "Banana Bonanza," *New Scientist* (July 21, 2001), 7.

[110] *Knight-Ridder/Tribune News Service*, (July 8, 1994).

[111] *Los Angeles Times*, (July 16, 1998), A-3.

[112] Richardson interview; *Los Angeles Times,* (Sept. 24, 1997), A-3, (Sept. 28, 1997, Ventura Edition), B-18, (Feb. 10, 1998, Ventura Edition), B-1, (March 1, 1998, Ventura Edition), B-16.

Chapter 3 Notes: Banana Business

[113] Charles Morrow Wilson. *Empire in Green and Gold: The Story of the American Banana Trade*. (New York: Henry Holt, 1947), 4-5.

[114] Tom Barry and Deb Preusch. *The Central America Fact Book*. (New York: Grove Press, 1986), 147-148.

[115] Clifford Krauss. *Inside Central American: Its People, Politics and History.* (New York: Summit Books, 1991), 183-184.

[116] Ernest Hamlin Baker,"United Fruit," *Fortune* (March 1933), 34-36.

[117] Baker, "United Fruit," 36; "United Fruit: 50,000,000 Bunches," *Newsweek* (March 30, 1935), 35.

[118] Quoted in Krauss, 186.

[119] Wilson, *Empire,* 24.

[120] Frederick Upham Adams, *Conquest of the Tropics.* (Garden City, N. Y., 1914), 35.

[121] Wilson, *Empire,*, 15.

[122] Adams, 16

[123] Wilson, *Empire*, 26-27

[124] Peter N. Davies, *Fyffes and the Banana: Musa Sapientum - A Centenary History 1888-1988.* (Atlantic Highlands, N. J., Athlone Press, 1990), 25-26.

[125] Charles Foster Batchelder, *Tropic Gold: the Story of the Banana Pioneer, Captain Lorenzo Dow Baker,* (Boston: W. D. Bradstreet, 1951), 2-3, 18.

[126] Wilson, *Empire,* 18-19; Charles Morrow Wilson, *Dow Baker and the Great Banana Fleet* (Harrisburg, Pa., Stackpole Books, 1972), 46.

[127] Batchelder, 20, 25; Wilson, *Empire,* 18.

[128] Wilson, *Empire,* 19-20; Wilson, *Empire*, 38-39.

[129] Wilson, *Empire*, 22-23; Batchelder, 34.

[130] Wilson, *Empire*, 28-29

[131] Quoted in Davies, 29.

[132] Davies, 28-30; Batchelder, 66.

[133] Davies 32; Wilson, *Empire,* 80-81.

[134] Adams, 70-72.

[135] Davies, 32-33; Baker, "United Fruit," 30; Mark H. Goldberg, *Going Bananas": 100 Years of American Fruit Ships in the Caribbean* (Kings Point, N.Y., Merchant Marine Museum Foundation, 1993), 251-252.

[136] Wilson, *Empire,* 83.

[137] Wilson, *Empire,* 75-7

[138] After 1890 over one hundred of the fruit boats were part of what was known as the "Mosquito Fleet,"ships that the banana companies chartered from Norwegian owners.

[139] homas L. Karnes, *Tropical Enterprise: The Standard Fruit and Steamship Co, in Latin America.* (Baton Rouge: LSU Press, 1978), 25-27.

[140] Watt Stewart, *Keith and Costa Rica: A Biographical Study of Minor Cooper Keith,* (Albuquerque, University of New Mexico Press, 1964), 19-21.

[141] Aviva Chomsky, *West Indian Workers and the United Fruit Company in Costa Rica, 1870-1940.* (Baton Rouge: L.S.U. Press, 1996), 17-18.

[142] Steward, 21, 34-35.

[143] Wilson, *Empire,* 46; Chomsky, 23.

[144] Wilson, 49-50.

[145] Chomsky, 24: Wilson, *Empire,* 49-50.

[146] Stewart, 67.

[147] Stewart, 43.

[148] Wilson, *Empire,* 49.

[149] Baker, "United Fruit," 30

[150] Stewart, 29-32, 59.

[151] John Keith Hatch, *Minor C. Keith, Pioneer of the American Tropics,* (McLean, Va: n.p., 1963), 30, 35.

[152] Stewart, 148-149, Hatch, 51-52.

[153] Stewart, 155; Hatch, 53-54; Davies, 34.

[154] Quoted in Stewart, 157.

[155] Davies, 34-35; Stewart, 158-159.

[156] Wilson, *Empire,* 110.

[157] Goldberg, 7, 254.

[158] Lester D. Langley and Thomas Schoonover, *The Banana Men: American Mercenaries and Entrepreneurs in Central America, 1880-1930,* (Lexington, University of Kentucky Press, 1995), 37.

[159] Stacy May and Galo Plaza, *The United Fruit Company in Latin America (U. S. Business Performance Abroad, vol. 7),* (National Planning Association, 1958), 13.

[160] Thomas Schoonover, *The United States in Central America, 1860-1911: Episodes of Social Imperialism and Imperial Rivalry in the World System,* (Durham: Duke University Press, 1991), 35.

[161] Karnes, 16-18.

[162] Paul J. Dosal. *Doing Business With the Dictators: A Political History of United Fruit in Guatemala, 1899-1944.* (Wilmington, DE, 1933), 5.

[163] Karnes, 2-3.

[164] Karnes, 7-8.

[165] Karnes, 47-49, 92; Wilson, *Empire,* 113.

[166] John Kobler, "Sam the Banana Man," *Life,* 30 (Feb. 19, 1951), 83-84.

[167] Kobler, 87.

[168] Kobler, 87; Thomas P. McCann, *An American Company: The Tragedy of United Fruit,* (New York: Crown, 1976), 18; Stephen J. Whitfield, "Strange Fruit: The Career of Samuel Zemurray," *American Jewish History.* 73 (March, 1984), 309; Henry F. Pringle, "A Jonah Who Swallowed the Whale," *American Magazine,* 116 (Sept. 1933), 114.

[169] Pringle 114; Kobler, 97.

[170] Barry, 253.

[171] Langley, *Banana Men,* 41-42; Krauss, 183; Barry, 253; Baker, "United Fruit," 32.

[172] Pringle, 114-115; Whitfield, 310.

[173] Baker,"United Fruit," 31.

[174] Kobler, 88.

[175] Lester D. Langley, *The Banana Wars: U. S. Intervention in the Caribbean, 1898-1934.* (Lexington: University of Kentucky Press, 1985), 177-178.

[176] Eugene Cunningham, *Triggernometry: A Gallery of Gunfighters* (Norman: University of Oklahoma Press, 1996), 389-390.

[177] Cunningham, 391-396.

[178] Cunningham, 398.

[179] Hermann B. Deutsch, *The Incredible Yanqui: the Career of Lee Christmas* (New York: Longmans, Green and Co., 1931), 74-77.

[180] Deutsch, 102-111.

[181] Baker, "United Fruit," 31.

[182] Deutsch, 111-115.

[183] Deutsch, 116-118.

[184] Karnes, 44; Langley, *Banana Men*, 136-138.

[185] Kobler, 92; Whitfield, 311-312.

[186] Whitfield, 312; Kobler, 91.

[187] Baker, "United Fruit," 28.

[188] Wilson, *Empire*, 248-249; May, 17; Kobler, 92; Whitfield, 312-313.

[189] McCann, 21; Kobler, 92; Baker, "United Fruit," 26; Whitfield, 313. The historical recounting of Zemurray's comment is in itself amusing. One writer chose to use the word "botched" in place of the stronger Anglo-Saxon verb which he said is never transferred to the pages of scholarly journals. Another writer in the preface to his book apologized to his mother for use of the vulgarity.

[190] Baker, "United Fruit," 29; McCann, 22; Whitfield, 313; Langley, *Banana Men*, 168.

[191] Kobler, 79; Whitfield, 315.

[192] Kobler, 82; *Newsweek_* (March 30, 1935), 34.

[193] McCann, 22; Kobler, 92-93.

[194] McCann, 25.

[195] Goldman, 417-418.

[196] Goldberg, 435.

[197] *Unifruitco*, (Nov. 1948), 3; Karnes, 210, 213.

[198] Edward L. Bernays, *Biography of an Idea: Memoirs of Public Relations Counsel Edward L. Bernays*, (New York: Simon and Schuster, 1965), 745-6; McCann, 24.

[199] John H. Melville, *The Great White Fleet*, (New York: Vantage Press, 1976), 34, 52, 56.

[200] Goldberg, 153-4; Karnes, 210.

[201] Quoted in Whitfield, 315.

[202] Bernays, 747.

[203] Chomsky, 110-113.

[204] McCann, 143.

[205] Whitfield, 315.

[206] Kobler, 94.

[207] *United Fruit Report*, (April 1951), 22.

[208] McCann, 39-40, 123.

[209] McCann, 26-27, 98.

Chapter 4 Notes: The Immigrant Banana

[210] Josephine Cirella. Interview with author, February 10, 1997. Palm Springs, California.

[211] Josephine Cirella. Interview #EI-202, August 15, 1992. Ellis Island Oral History Project.

[212] Cirella. Interview with author. Feb. 10, 1997; Ellis Island Interview #202.

[213] Sophie Shuman Taub. Interview #KM-39, April 25, 1994. Ellis Island Oral History Project.

[214] Dr. Janet Levine. Interview with author. Ellis Island, New York, Dec. 9, 1996.

[215] Angela Garofalo Basso. Interview #E-186, June 27, 1992. Ellis Island Oral History Project.

[216] Ellis Island interview #202.

[217] Oreste Teglia. Interview #AKRF-105, Dec. 20, 1985. Ellis Island Oral History Project.

[218] Patrick Peak. Interview #AKRF-84, Nov. 15, 1985. Ellis Island Oral History Project.

[219] Ellis Island Interview #AKRF-84.

[220] Sonya Thornblom Gillick. Interview #EI-302, April 27, 1993. Ellis Island Oral History Project.

[221] Anna Kikta. Interview#NPS-160, May 11, 1988. Ellis Island Oral History Project.

[222] Louis V. Stoller. Interview #EI-4, September 5, 1990. Ellis Island Oral History Project.

[223] Lawrence Meinwald. Interview #50, May 29, 1991. Ellis Island Oral History Project.

[224] Czeslawa Palenska Lutz. Interview #EI-142, April 28, 1992. Ellis Island Oral History Project.

[225] Madeline Polignano Zambrano. Interview EI-76, August 27, 1991. Ellis Island Oral History Project.

[226] Hilda and Arthur Broskas. Interview #EI-157, May 23, 1992. Ellis Island Oral History Project.

[227] Tessie (Anastasia} Kalogarados Argianas. Interview #AKRF-106, Dec. 19,

1985. Ellis Island Oral History Project.

[228] Sara Milanow Rovner. Interview #EI-104, Oct. 4, 1991. Ellis Island Oral History Project.

[229] Rovner, Interview #EI-104.

[230] Rovner. Interview #EI-104.

[231] Ethel Domerschick Pine. Interview #EI-144, May 7, 1992. Ellis Island Oral History Project.

[232] George Monezis. Interview #EI-413, May 28, 1989. Ellis Island Oral History Project.

. [233] Rocco Morelli. Interview #AKRF-31, Sept. 16, 1985. Ellis Island Oral History Project.

[234] Salvatore Fichera. Interview #EI-165, June 3, 1992. Ellis Island Oral History Project.

[235] Anna Tonnesen Bjorland. Interview EI-355, July 21, 1993. Ellis Island Oral History Project.

[236] Sarah Asher Crespi. Interview #EI-29, September 1, 1993. Ellis Island Oral History Project. The interview was on the day Mrs. Crespi celebrated her ninetieth birthday.

[237] Elizabeth Milovsky Friedman. Interview #EI-423, Nov. 3, 1989. Ellis Island Oral History Project.

[238] Euterpe Boukis Dukakis. Interview #AKRF-91, Nov. 21, 1985. Ellis Island Oral History Project.

[239] Jack Avruch. Interview #EI-71, August 24, 1991. Ellis Island Oral History Project.

[240] Verna Trill Horvath. Interview #EI-400, May 23, 1989. Ellis Island Oral History Project.

[241] Elizabeth Repshis. Interview #AKRF-46, October 11, 1985— Ellis Island Oral History Project.

[242] Esther Shapiro Gologar. Interview EI-143, April 29, 1992. Ellis Island Oral History Project.

[243] Marie Tancibok Vitosky. Interview EI-249, February 19, 1993. Ellis Island Oral History Project.

[244] Mary Masare Thome. Interview #DP-54, November 6, 1989. Ellis Island Oral History Project.

[245] Charles Morrow Wilson, *Empire in Green and Gold: The Story of the American Banana Trade* (New York: Henry Holt, 1947), p. 87

[246] Marianthe Dimitri Chletsos. Interview #AKRF-86, —. Ellis Island Oral History Project.

[247] Rose Ganbaum Halpern. Interview #EI-306, April 28, 1993. Ellis Island Oral History Project.

[248] Pauline Stevens Curtis. Interview #EI-215, September 18, 1992. Ellis Island Oral History Project.

[249] John Santo Fieramosca. Interview #EI-471, May 4, 1994. Ellis Island Oral History Project.

[250] Betty Dornbaum Schubert. Interview #EI-9, November 7, 1990. Ellis Island Oral History Project.

[251] Roslyn Bresnick-Perry. Interview #EI-482, June 30, 1994. Ellis Island Oral History Project.

[252] Cirella. Ellis Island Interview #EI-202.

[253] Gillick. Ellis Island Interview #EI-302.

Chapter 5 Notes: Banana Republic Politics

[254] Clifford Krauss, *Inside Central America: Its People, Politics and History*, (New York: Summit Books, 1991), 178.

[255] Stephen Schlesinger and Stephen Kinzer, *Bitter Fruit: The Untold Story of the American Coup in Guatemala*, (New York: Doubleday/Anchor, 1983), 7-9.

[256] T. Harry Williams, *Huey Long*, (New York: Alfred A. Knopf, 1970), 466-468.

[257] Schlesinger, Fruit, 25-35; Alejandra Batres, *The Experience of the Guatemalan United Fruit Company Workers, 1944-1954: Why Did They Fail?*, Texas Papers on Latin America, Paper No. 95-01, (Austin: University of Texas, 1995), 1.

[258] Schlesinger, *Fruit*, 50; Batres, 23; Don Moore, *The Clandestine Granddaddy of Central America*, available from Moore@acc.mcrest.edu, 1989.

[259] Schlesinger, *Fruit*, 75-76.

[260] Thomas McCann, *An American Company: The Tragedy of United Fruit*, (New York: Basic Books, 1996), 47.

[261] E. Howard Hunt, *Undercover: Memoirs of an American Secret Agent*, New York: Berkeley Publishing Corp., 1974), 97.

[262] Schlesinger, *Fruit*, 58-63.

[263] Jon Lee Anderson, *Ché Guevara: A Revolutionary Life*, (New York: Grove Press, 1997), 126, 176.

[264] Hunt, 99.

[265] Stephen Schlesinger , "The C.I.A. Censors History," *The Nation*, (July 14, 1997), 20. The plans for assassination were unknown to the public until the CIA released on May 23, 1997 about 1,400 of 180,000 pages of classified documents on the Guatemalan operation.

[266] Jim Schrider, "CIA Coup Files Include Assassination Manual," *National Catholic Reporter*, (July 18, 1997), 10.

[267] Burton Hersh, *The Old Boys: The American Elite and the Origins of the CIA*, (New York: Scribner's, 1992), 343.

[268] Hunt, 98.

[269] Schlesinger, *Fruit*, 114-116, 119-122; Moore, 4.

[270] Moore , 4-5.

[271] Schlesinger, *Fruit*, 124.

[272] Moore, 5-9; Schlesinger, *Fruit*, 167-169.

[273] Hersh, 347.

[274] Schlesinger, *Fruit*, 179-182.

[275] Schlesinger, *Fruit*, 187; Moore, 9-10.

[276] Schlesinger, *Fruit*, 205-225.

[277] Schlesinger, *Fruit*, 193-194. The CIA later compensated Lloyd's of London $1.5 million for the loss.

[278] Schlesinger, *Fruit*, 220-221; McCann, 62, 67.

[279] McCann, 93.

[280] *New York Times*, (Sept. 20, 1996), A8; (Jan. 5, 1997), 4:1.

[281] *Time*_, (Sept. 3, 1973), 76; Eleanor Johnson Tracy, "How United Brands Survived the Banana War," *Fortune*, (July 1976), 146; *New York Times*, (Feb. 4, 1975), 1.

[282] *New York Times*, (Feb. 4, 1975), 10.

[283] *New York Times*, (Feb. 4, 1975), 10.

[284] Tom Barry and Deb Preusch, *The Central American Fact Book*, (New York: Grove Press, 1986), 151-152; *Newsweek*, (April 21, 1975), 79-81.

[285] *Business Week*, (Feb. 14, 1977), 37; (Sept. 19, 1977), 40.

[286] James Ring Adams, *The Big Fix: Inside the S & L Scandal*, (New York: John Wiley & Sons, Inc., 1990), 238; "So You Want to Buy a President" available at Frontline/ WGBH Educational Foundation/ www.wgbh.org

[287] Johnson Tracy, 145.

[288] Jeff Harrington, "A Soldier in the Banana War," *Cincinnati Enquirer*, (Dec. 10, 1995), H1; "The Banana Rebellion," *Time*, (June 11, 1990), 49; Kent Norsworthy and Tom Barry, *Inside Honduras*, (Albuquerque, N. M.: Inter-Hemispheric Education Resources Center, 1994), 70-71.

[289] Harrington, "Soldier," H1.

[290] John Otis, "Bruised Bananas," *Latin Trade*, (February 1997), 28-30.

[291] Otis, 30; Anders Corr, "The Chiquita Republic," 1997 available at E-News Online.

[292] *Cancer Weekly Plus*, (May 19, 1997), 20.

[293] Jeff Harrington, "Pesticide Legacy Nags Chiquita," *Cincinnati Enquirer*, (Aug. 27, 1995), G1.

[294] Harrington, "Pesticide," G1.

[295] Tom Boswell, "Begging for Peace Amid so Much Death," *National Catholic Reporter*, (Jan. 24, 1997), 9.

[296] Joshua Hammer, "Lining Up at the Exit," *Newsweek*, (March 6, 1995), 38.

Chapter 6 Notes: Selling Bananas

[297] Herman van Beek, "The Importance of Social and Environmental Clausing for A Sustainable Banana Industry" (paper presented at the International Banana Conference, "Toward a Sustainable Banana Economy," Brussels, Belgium, May 4-6, 1998), 1.

[298] International Banana Association, *Banana Facts from 2001 Consumer Research*, (April 3, 2002).

[299] U. S. Department of Agriculture, *Situation and Outlook: Fruit and Tree Nuts*, (March 30, 1998).

[300] Phil Lempert, "Increasingly Exotic Produce Mystifies Many U. S. Shoppers," *The (Newark, N. J.) Star Ledger*, (Dec. 3, 1997), 36.

[301] *The Packer, Fresh Trends*, (1997).

[302] *Progressive Grocer, Supermarket Business*, (Sept., 1997).

[303] Sherrie Terry, Vice-President for Marketing, Chiquita Banana North America, interview with the author, Cincinnati, Ohio, May 28, 1998.

[304] *Unifruitco*, (Oct. 1953).

[305] *New York Times*, (Oct. 24, 1954), 1.

[306] *New York Times*, (Dec. 9, 1954), 51.

[307] *Fruit Dispatch*, (Aug. 1925), 169.

[308] Thomas L. Karnes, *Tropical; Enterprise: The Standard Fruit and Steamship Co. in Latin America*, (Baton Rouge: LSU Press, 1978), 284-285.

[309] Robert A. Read, "The Banana Industry: Oligopoly and Barriers to Entry" in Casson, Mark. *Multinationals and World Trade: Vertical Integration and the Division of Labour in World Industries.* (London: Allen and Unwin, 1986).

[310] Thomas P. McCann, *An American Company: The Tragedy of United Fruit*, (New York: Crown Publishers, 1976), 71-72.

[311] S. Terry, Chiquita Banana, interview with author, May 28, 1998.

[312] *Supermarket Business* (April 15, 2001), 96.

[313] Karnes, 30.

[314] *Unifruitco*, (Oct. 1948), 4.

[315] *New York Times*, (Aug. 6, 1956), 45; (Nov. 28, 1952), 43; (Dec. 21, 1952), 10; (March 8, 1953), 10.

[316] *The (Newark, N. J.) Star Ledger*, (Jan. 22, 1988); (Sept. 23, 1997), 41-42; (Jan. 9, 1998), 41.

[317] *Baltimore Sun*, (Feb. 11, 1990).

[318] Charles Morrow Wilson, *Empire in Green and Gold: The Story of the American Banana Trade*, (Henry Holt, 1947), 99.

[319] "Fifty Years of Fruit Dispatch," *Unifruitco*, (Dec. 1948), 14.

[320] Quoted in Wilson, 180.

[321] *Unifruitco*, (Oct. 1948): 6; *United Fruit Report*, (July 1951), 1-2, 6.

[322] *Fruit Dispatch*, (Nov. 1917), 208-209.

[323] *Supermarket News*, (Aug. 2, 1993), 34.

[324] *American Shipper*, (April 1995), 48.

[325] D. K. Salunkhe, H. R. Bolin and N. R. Reddy, *Storage, Processing and Nutritional Quality of Fruits and Vegetables*, 2nd ed., Vol 1, Fresh Fruits and Vegetables, (Boca Raton, Florida: CRC Press, 1991), 54.

[326] *Unifruitco*, (Dec. 1948), 18; *New York Times*, (June 14, 1949), 35.

[327] *Fruit Dispatch*, (May 1917), 12.

[328] Rick Sebak, producer, *The Strip Show*, (WQED, Pittsburgh, 1997).

[329] *Your Guide to Greater Profits*, Chiquita Brands International Inc.

[330] Salunkhe, 60.

[331] Andrew Kaplan, "Banana Ripening Reefers Hit the Highway," *Food Logistics* (Jan-Feb. 2002), 8.

[332] Quoted in *Fruit Dispatch*, (Jan. 1918), 265.

[333] Elizabeth Somers, M.D., *Essential Guide to Vitamins and Minerals*, (New York:HarperCollins, 1992), 18, 125, 129.

[334] Frederick Upham Adams, *Conquest of the Tropics*, (Garden City, N. Y., 1914), 338.

[335] U S Department of Agriculture, (March 2002).

[336] Karnes, 28.

[337] Wilson, 189.

[338] Adams, 334-335; Karnes, 52.

[339] *Fruit Dispatch*, (May 1917), 25.

[340] *Fruit Dispatch*, (Jan. 1918), 318.

[341] *Fruit Dispatch*, (May 1917), 20-21.

[342] *Fruit Dispatch*, (July 1917), 81.

[343] *Fruit Dispatch*, (May 1917), 19.

[344] Wilson, 228.

[345] *Fruit Dispatch*, (April 1926), 480.

[346] *Good Housekeeping*, (August 1926), 82-83.

[347] Benjamin Spock and Mary Morgan, *Spock on Spock: A Memoir of Growing Up With the Century*, (New York: Pantheon Books, 1989), 24-25.

[348] *Fruit Dispatch*, (Jan. 1926), 379-382

[349] *Fruit Dispatch*, (Jan. 1926): 361-362; (July 1926), 143-146.

[350] *Good Housekeeping*, (July 1926), 197.

[351] *Good Housekeeping*, (Sept. 1926), 26+

[352] Karnes, 177; Read, 319.

[353] *Fruit Dispatch*, (June 1926), 98; (June, 1929), 6-7.

[354] *Fruit Dispatch*, (June 1926), 113; (July, 1926), 140-141; (Aug. 1926), 224.

[355] *Fruit Dispatch*, (June 1926), 75.

[356] *Fruit Dispatch*, (Nov. 1927), 241; (Feb. 1929), 5.

[357] *Fruit Dispatch*, (May 1917), 29.

[358] *Unifruitco*, (Feb. 1930), 404-406.

[359] *Unifruitco*, (March 1930), 465.

[360] *Unifruitco*, (March 1930), 466.

[361] *Unifruitco*, (Feb 1931), 356-358.

[362] Association of American Railroads, *Bananas, tropical, n.o.s. dried or evaporated fruits*, (Washington: Association of American Railroads, 1946), 27 and 46.

[363] Wilson, 289; *New York Times*, (May 28, 1946), 18.

[364] *Unifruitco*, (Nov. 1948), 12 -14.

[365] *New York Times*, (April 20, 1944), 25.

[366] *New York Times*, (May 6, 1946), 121; *Unifruitco*, (Feb. 1948), 3.

[367] *Unifruitco*, (Oct. 1948), 25.

[368] *New York Times*, (Sept. 5, 1946), 29; (May 13, 1947), 29, 43.

[369] *New York Times*, (Oct. 11, 1946), 18; (Nov. 16, 1948), 1; (Oct. 25, 1951), 24.

[370] *Fruit Dispatch*, (Jan.-Feb. 1925), 442.

[371] Wilson, 184 - 185; Chiquita Brands International, Inc.

[372] Chiquita Brands International, Inc.

[373] *Unifruitco*, (Nov. 1948), 19.

[374] *Wall Street Journal*, (May 5, 1994), B1.

[375] McCann, 73-75.

[376] McCann, 75.

[377] The Gallup Organization, 1997.

[378] Equitrend, 1998.

[379] McCann, 89.

[380] *Newark (N.J.)Star-Ledger*, (May 1, 1994).

[381] McCann, 87.

Chapter 7 Notes: A Banana Eat Banana World

[382] "Sweet Success," *Sarasota Herald Tribune*, (June 11, 2001), 12.

[383] Karnes, *Tropical Enterprise: The Standard Fruit and Steamship Co. in Latin America*, (Baton Rouge: LSU Press, 1978), 286, 289, 293.

[384] Karnes, 277, 293-294.

[385] Tom Barry and Deb Preusch, *The Central American Fact Book*, (New York: Grove Press, 1986), 148-149.

[386] Dole Food Company, Inc., *2000 Annual Report*; Chiquita Brands International, Inc., *2001 Annual Report*.

[387] Fresh Del Monte Produce, Inc., *1997 Annual Report*; *Supermarket News*, (April 20, 1998), 73.

[388] Vince Staten, *Can You Trust a Tomato in January?* (New York: Simon & Schuster, 1993), 32, 39.

[389] Fresh Del Monte Produce

[390] Point-of-Purchase Advertising Institute, *Consumer Buying Habits Study*, 1986.

[391] Fresh Del Monte Produce

[392] Chiquita Banana North America; Fresh Del Monte Produce; Turbana Corporation.

[393] *Supermarket News*, (Oct. 7, 1996), 48.

[394] *Supermarket News*, (April 21, 1997), 69.

[395] *Supermarket News* (Nov. 22, 1999), 24

[396] *Wall Street Journal*, (Nov. 24, 1997), B14A.

[397] Dole Food Company ; http//www.dole.com

[398] S. Terry, Chiquita Banana, interview with author, May 28, 1998.

[399] Laura Bird, "Chiquita's Ad Archive: The Picture of Health," *AdWeek's Marketing Week*, (Jan. 7, 1991), 32-33.

[400] *New York Times*, (March 8, 1993), D8.

[401] W. B. Doner , *Chiquita: Historical Reel, 1940s to Present*, (Multi-Image Productions, 1998), videotape.

[402] *Supermarket News*, (Sept. 18, 1995), 24.

[403] Fara Warner, "Surprise! Chiquita Advises Eat Bananas," *AdWeek's Marketing Week*, (April 29, 1991), 8.

[404] *Supermarket News*, (Oct. 19, 1992), 34.

[405] S. Terry, Chiquita Banana, interview with author, May 28, 1998.

[406] Herman van Beek, "The Role of Fair Trade and the Social and Environmental Responsibility of the Private Sector," (paper presented at the International Banana Conference, "Toward a Sustainable Banana Economy," Brussels, Belgium, May 4-6, 1998), 1-3; *Banana Trade News Bulletin*, (Feb. 1997); (July 1997).

[407] *Banana Trade News Bulletin*, (July 1997); van Beek, "Role of Fair Trade."

[408] "First-Time Report from Chiquita impresses on Every Count," *Business and the Environment*, (Nov. 2001), 5.

[409] "Better Bananas," *Environment*, (Jan. 2001), 4.

[410] Chiquita Brands International, Inc. *2000 Corporate Responsibility Report*, 2001.

[411] available at www.dole.com, July 2002.

[412] "Dole Grows Organic Bananas for Sale in North America," *Business and the Environment*, (Feb. 2001), 8; both organic and traditional bananas were found to be selling at sixty-nine cents a pound in a New Jersey supermarket in July 2002.

[413] available at www.freshdelmonte.com

[414] Magnes Welsh, Chiquita Brands International, Inc., interview with the author, May 28, 1998.

Chapter 8 Notes: The Great Banana War

[415] Steve Rios, "Europe's Forbidden Fruit," *Hispanic,* (July 1993), 12.

[416] Lawrence S. Grossman, *The Political Ecology of Bananas: Contract Farming, Peasants, and Agrarian Change in the Eastern Caribbean,* (Chapel Hill, NC: University of North Carolina Press, 1998).

[417] John Rodden, "Going Bananas: A Story of Forbidden Fruit," *Commonweal,* (March 24, 1995), 5.

[418] "Bananas Make Berlin Glad," *Fruit Dispatch,* (Nov. 1919), 244.

[419] *New York Times,* (Sept. 12, 1940), 7.

[420] Rodden, 5; "Yes, We Have No Bananas," *Canada and the World Backgrounder,* (Sept. 1994), 18.

[421] John Rothchild, "The Banana Wars," *Time,* (Oct. 17, 1994), 39; Rodden, 5-6.

[422] *Journal of Commerce and Commercial,* (July 19, 1995), 1A-2A.

[423] *Journal of Commerce and Commercial,* 1A-2A.

[424] John Otis, "Bruised Bananas," *Latin Trade,* (February 1997), 29.

[425] *Environmental Action,* (Winter, 1996), 2.

[426] Quoted in *Environmental Action,* (Winter, 1996), 2.

[427] Brook Larmer, "Brawl Over Bananas," *Newsweek,* (April 28, 1997), 43.

[428] *New York Times,* (Dec. 5, 1995), B9. The reference is to Doug Richardson's Seaside Banana Gardens.

[429] *Washington Post,* (Feb. 26, 1995), A18.

[430] *New York Times,* (Dec. 5, 1995), A1; John Greenwald, "Banana Republican," *Time,* (Jan. 22, 1996), 54-55; *Journal of Commerce and Commercial,* (Jan. 11, 1996), 1A.

[431] Leslie Wirpsa, "Payoff maybe tied to Drug War Hitch," *National Catholic Reporter,* (April 5, 1996), 16.

[432] Michael Weisskopf, "The Busy Back-Door Men," *Time,* (March 31, 1997), 40; William Safire, "Bananagate," *New York Times,* (March 26, 1997), A21; Larmer, 43.

[433] *Washington Post,* (June 6, 1996), A27; Bob Herbert, "Banana Bully," *New*

York Times, (May 13, 1996), A15; *Jet*, (Jan. 20, 1997), 47.

[434] Quoted in *Washington Post*, (March 19, 1997), C9.

[435] *New York Times* , (May 9, 1997), A6.

[436] Larmer, 43.

[437] *The Economist*¸ (May 31, 1997), 36; *New York Times*, (May 11, 1997), H2.

[438] *Christian Science Monitor*, (Nov. 13, 1997), 7; *The Economist*, (May 31, 1997), 36.

[439] "Expelled from Eden," *The Economist*, (Dec. 20, 1997), 35-38.

[440] "Banana Split," *Mother Jones*, (Nov.-Dec, 1998), 62.

[441] "Monkey Business," *The Economist*, (Nov. 14, 1998), 82.

[442] David E. Sanger, "Unreal Aura for Routine Business in Washington and the White House," *New York Times*, (Dec. 19, 1998).

[443] *Time* (Feb. 7, 2000), 52.

[444] "A Fruity Peace," *The Economist* (April 21, 2001), 2.

[445] "Banana Workers in Panama Strike to Protest Wages, Working Conditions," *Miami Herald*, (Sept. 16, 2001).

[446] "New Ideas that May Bear Fruit," *The Financial Times,* (Nov. 17, 2000), 40.

[447] "Shady Crop: Child Workers Exploited in Ecuador Banana Plantations," *EFE World News Service,* (April 24, 2002).

[448] Juan Forero, "In Ecuador's Banana Fields, Child Labor is Key to Profits," *The New York Times*, (July 13, 2002), 1, 6.

[449] See www.dole.com, 2002

[450] *2000 Corporate Responsibility Report, (*Chiquita Brands International, Inc., 2001), 39.

[451] "Ecuador Wants to Fight Child-labor Abuse on Banana Plantations," *EFE World News Service,* (June 12, 2002); Forero, 6.

[452] Nicholas Stein, "Yes, We Have No Profits: The Rise and Fall of Chiquita Banana: How a Great American Brand Lost its Way," *Fortune,* Nov. 26, 2001), 182+.

[453] Stein.

[454] Edward Alden and Christopher Bowe, "US Banana Producers Find Different Ways of Living with EU Restrictions," *The Financial Times,* (Feb. 16, 2001), 12.

Chapter 9 Notes: The Healthy Banana

[455] Anonymous, *Fruit Dispatch*, (May 1923), 24.

[456] *New York Times*, (Sept. 9, 1996); *Newark (N.J.)Star Ledger*, (Sept. 8, 1996), Sec. 5, 19.; Brian Cleary, "Playing to the Crowd," *Tennis*, (Nov. 1996), 62, 66.

[457] *Unifruitco*, (Oct. 1930), 152.

[458] "Bananarama," *Runner's World*, (Oct. 1999), 16.

[459] "Real Men Eat Broccoli—and Bananas and Oranges," *Tufts University Diet & Nutrition Letter*, (Nov. 1993), 8.

[460] D.K. Salunke, H. R. Bolin and N. R. Reddy, *Storage, Processing and Nutritional Quality of Fruits and Vegetables*, 2nd ed. Vol. I, Fresh Fruits and Vegetables, (Boca Raton, FL: CRC Press, 1991), 24.

[461] Joe Lindsey, "Cycling's Perfect Food, " *Bicycling*,(May 2002), 28.

[462] Bruce Adams, "49ers Hydrate to Avoid Being Left High and Dry," *San Francisco Examiner*, (Aug. 30, 1997) , B1; available at http://www.sfgate.com.

[463] Sheldon Margen, M.D. and Dale A. Ogar, "Consider the Kiwi," *Newark (N.J.) Star Ledger*, (Oct. 5, 1997), 2:8.

[464] *Fruit Dispatch*, (July 1919), 3

[465] *The New York Times*, (Sept. 20, 1952), 14.

[466] P. K. Reynolds and C. Y. Yang, "The Banana in Chinese Literature," *Harvard Journal of Asiatic Studies*, (Vol. 5), 170-177.

[467] Richard A. Shweder, "Ancient Cures for Open Minds," *New York Times*, (Oct. 26, 1997), 4:6.

[468] Jean Carper, *The Food Pharmacy*, (New York: Bantam, 1988), 122.

[469] Lelord Kordel, *Natural Folk Remedies*, (New York: Putnam, 1974), 99.

[470] A. T. Bannon, "How Bananas Helped Explorer Stanley," *Fruit Dispatch*, (July 1918), 83-84.

[471] Ann Landers, *Newark (N.J.) Sunday Star Ledger*, (Nov. 30, 1997), 2:11.

[472] *Fruit Dispatch*, (March 1929), 11.

[473] Sidney V. Haas, M.D., "The Value of the Banana in the Treatment of Celiac Disease," *American Journal of Diseases of Children*, (Oct. 1924) quoted in *Fruit Dispatch*, (August 1925), 167.

[474] *Fruit Dispatch*, (June 1931), 560.

[475] *Unifruitco*, (May 1954), 3.

[476] "Banana Priorities," *Newsweek*, (Aug. 10, 1942), 57-58.

[477] *New York Times*, (Jan. 19, 1947), 1; (Jan. 20, 1947), 6; (Jan. 21, 1947), 21; (Jan. 22, 1947), 26.

[478] *New York Times*, (June 16, 1950), 27.

[479] Jennifer Quasha, "BRATT: Be Nice to Your Tummy Diet," *Total Health*, (June 1995), 20.

[480] Ronald Hoffman, M.D., *Seven Weeks to a Settled Stomach*, quoted in Jean Carper, *The Food Pharmacy*, (New York: Bantam, 1988), 133.

[481] Carper, *Food Pharmacy*, 124; Jean Carper, *Food - Your Miracle Medicine*, (New York: HarperCollins, 1993), 175-176; Editors of Prevention, *Complete Book of Natural and Medicinal Cures*, (Rodale Press, 1994), 37.

[482] "A Banana a Day," *Reader's Digest*, (March 1993), 46; *Complete Book of Natural and Medicinal Cures*, 36-37.

[483] *New York Times*, (Sept. 22, 1998), F-7.

[484] "Up With Potassium, Down With Blood Pressure," *Tufts University Health & Nutrition Letter*, (Aug. 1997), 7.

[485] Carper, *Food Pharmacy*, 124.

[486] Charles Bixler Heiser, *Seed to Civilization: The Story of Food*, (Cambridge, Mass.: Harvard University Press, 1990), 156.

[487] "British Boy Faces Life Diet of Broccoli, Bananas" Reuters (Jan. 19, 1998) available at Infoseek: The News Channel" at http://www.infoseek.com; Research in Spain indicates a connection between latex allergy and allergy to bananas and chestnuts. The study was reported in *JAMA: The Journal of the American Medical Association*, (June 23, 1993), 3083.

[488] "The Smell of Weight Loss," *Food and Beverage Marketing*, (Jan. 1994), 37; "At Last the Perfect Decaf," *Food Processing*, (May 1995), 19.

[489] "Sniffing Your Way to Weight Loss? Something Smells Fishy Here," *Tufts University Health and Nutrition Letter*, (April 1998), 6.

[490] Scott Cunningham, *The Magic in Food: Legends, Lore and Spellwork*, (St. Paul, Minn.: Llewellyn Publications, 1991), 115.

[491] Hampton Sides, *Ghost Soldiers: The Epic Account of World War II's Greatest Rescue Mission,"* (New York: Anchor Books, 2001), 144.

[492] "Vaccine Cuisine Could be the Wave of the Future," *Infectious Diseases in Children*, (Jan. 22, 1997) available at http://www.slacking.com; National Institute of Environmental Health Sciences, "Vaccine Cuisine," *Environmental Health Perspectives*, (March 1996) available at http://

ehpnet1.niehs.nih.gov; "Dishing Up a Mess O'Immunity," *U. S. News and World Report*, (April 24, 1995), 16.

[493] William H. R. Langridge, *Scientific American*, (Sept. 2000).

[494] Eryn Brown, "A Banana a Day...," *Fortune*, (May 2001).

[495] Charles Morrow Wilson, *Empire in Green and Gold: The Story of the American Banana Trade*, (New York: Henry Holt, 1947), 180.

[496] *Where's Thomas Pynchon*, (June 5, 1997), available from http://www.cnn.com.

[497] Thomas Moore, *The Style of Connectedness: Gravity's Rainbow and Thomas Pynchon*, Columbia: University of Missouri Press, 1987), 5.

[498] Moore, 1-5; *New York Times*, (May 8, 1974), 38. Pynchon refused the William Dean Howells Medal of the National Institute of Arts and Letters in 1975. He did note that it might be of some value because of its gold content.

[499] Pynchon may not have done his homework relative to growing bananas. It is unlikely that one would be picking ripe bananas from the tree. Most bananas that ripen on the tree have likely split their skins and become unappetizing.

[500] Thomas Pynchon, *Gravity's Rainbow*, (New York: Viking Press, 1973), 10.

[501] Sanford Ames, "Fast Food/Quick Lunch: Crews, Burroughs and Pynchon," in *Literary Gastronomy*, ed. David Bevan, (Amsterdam, Netherlands, Rodopi, 1988), 24.

[502] Pynchon, *Gravity's Rainbow*, 678.

[503] Quoted in Brigid Allen, ed., *Food: An Oxford Anthology*, (New York: Oxford University Press, 1994), 20-21.

[504] *Fruit Dispatch*, (Oct. 1917), 170.

[505] Jane and Michail Stern, *Elvis World*, (New York: Alfred A. Knopf, 1987), n.p.

[506] "The King's Eating Habits Deadly, Expert Says," CNN, (Aug. 16, 1995), available at http://www.cnn.com.

[507] Charles Nevin, "Elvis by Numbers," *The Guardian*, (July 18, 1997), T2.

[508] *Newark (N.J.) Star Ledger*, May 16, 1996), 73.

[509] Stern, n.p.

[510] "You Ain't Nuthin' But An Egghead," ABC News available at http://www.ABCNews.com; "Elvis Is Alive and Well at the University of Chicago," (Associated Press, Jan. 10, 1998) available at http:www.cnn.com.

[511] John DeMers, "Born in the Vieux Carre," *New Orleans Magazine*, (June

1996), 23.

[512] Amy Zuber, "Buy, Buy This American Pie: Sweet Treat is a Slice of the Upper Crust," *Nation's Restaurant News*, (Sept. 30, 1996), 55.

[513] Diane Toops, "Going Bananas: Chiquita Debuts Appealing Fat Replacement System," *Food Processing*, (Aug. 1997), 51.

[514] Cunningham, 114-115, 271, 308.

[515] Cunningham, 259, 301.

[516] Cunningham, 307.]

[517] Cynthia Mervis Watson, M.D., *Love Potions*, (Los Angeles: Putnam, 1993), 59-61.

[518] Jennifer Kornreich, "Sex on the Weekend, " MSNBC, available at http://www.msnbc.com

[519] *Fruit Dispatch*, (April 1919), 421.

Chapter 10 Notes: The Banana on Stage and Screen

[520] Charles and Louise Samuels. *Once Upon a Stage*. (New York: Dodd, Mead & Co., 1974), 161.

[521] *Scribner's Magazine* 30 (September 1901), 351-5 quoted in *American Vaudeville As Seen By Its Contemporaries*, ed. Charles W. Stein, (New York: Alfred A. Knopf, 1984), 40.

[522] Anthony Slide, *The Encyclopedia of Vaudeville*, (Westport, Conn.: Greenwood Press, 1994), 22.

[523] Slide, 285-6.

[524] Samuels, 40.

[525] Slide, 287.

[526] Slide, 426.

[527] Joel Sayre, "Meanwhile, What of Burlesque? Let 'Sliding Billy' Watson Tell It," *New York Herald Tribune* , (April 20, 1930), n.p., Billy Rose Theater Collection, New York Public Library Performing Arts Library.

[528] *Fruit Dispatch*, (Oct.1919), 217; In 1930 "Sliding Billy" offered a different explanation for his name. He said he arrived on stage late for a cue. He ran out, slipped and landed in a box. "It wowed 'em out front," he said.

[529] "Sliding Billy" Watson Clippings file at Billy Rose Theater Collection, New York Public Library for the Performing Arts.

[530] Morton Minsky and Milt Macklin, *Minsky's Burlesque*, (New York: Arbor

House, 1986), 129.

[531] Jean-Claude Baker, Josephine's adopted son, believes that Josephine's father was a white man. He says that Josephine's mother worked for a white German family in St. Louis. Josephine was born in the white "Female Hospital" and was there six weeks. Josephine's birth was registered, an event unusual for black births at the time. See Jean-Claude Baker and Chris Chase, *Josephine: The Hungry Heart*, (New York: Random House, 1993), 16-17.

[532] William Wiser, *The Great Good Place: American Expatriate Women in Paris*, (New York: William Norton, 1991), 276.

[533] Baker, 30.

[534] Some sources say she was pregnant, see Wiser, 275. Josephine's sister Margaret said she was never pregnant at age 13 with Willie; see Phyllis Rose, *Jazz Cleopatra: Josephine Baker in Her Time*, (New York: Doubleday, 1989), 37.

[535] Wiser, 276-277.

[536] Rose, 48-49.

[537] Lynn Haney, *Naked at the Feast: A Biography of Josephine Baker*, (New York: Dodd, Mead & Co., 1981), 30.

[538] Wiser, 274-278.

[539] Wiser , 274-275, 278.

[540] *New York Times*, (May 2, 1924), 2.

[541] Rose, 60-63.

[542] Quoted in Haney , 44.

[543] Wiser, 268; Charles Rearicks, *The French in Love and War: Popular Culture in the Era of the World Wars*, (New Haven: Yale University Press, 1997), 81.

[544] Wiser, 269.

[545] Baker, 5.

[546] Wiser, 270-271; Rose, 20-21.

[547] Rearicks, 81.

[548] Quoted in Wiser, 281.

[549] Rose, 29.

[550] Wiser, 272.

[551] Quoted in Rose, 32.

[552] Rose, 80.

[553] Rearicks, 82.

[554] Haney, 85, 89.

[555] Rose, 90-92.

[556] Quoted in Baker, 135.

[557] Wiser, 286-287; Haney, 99; Rose, 97-99.

[558] Baker, 135.

[559] Haney, 101-102; Wiser, 285; Baker, 144.

[560] Baker, 135.

[561] Quoted in Haney ,99.

[562] Haney, 104.

[563] Wiser, 293; Rose, 114.

[564] Quoted in Haney, 127.

[565] Rose, 128; Wiser, 296-298.

[566] Wiser, 298; Rose, 139-140.

[567] Wiser, 299.

[568] Quoted in Rose, 140-141.

[569] Wiser, 300-303; Haney, 161-163.

[570] Wiser, 286.

[571] Wiser, 303-304.

[572] Haney, 192; Wiser, 305.

[573] *New York Times* (Jan. 31, 1936), 17.

[574] Wiser , 307-309.

[575] Rearicks, 245, 248; Haney, 216; Wiser, 312-313.

[576] Stephen Papich, *Remembering Josephine: A Biography of Josephine Baker*, (Indianapolis: Bobbs-Merrill, 1976), 125-130.

[577] Rose, 195-196; Wiser, 313.

[578] Haney, 247-249, 251-253, 259.

[579] Rose, 241-242.

[580] Quoted in Baker, 326.

[581] Baker, 406, 364-365.

[582] Quoted in Baker, 474.

[583] Haney, 102.

[584] *Village Voice* (Oct. 29, 1991), 43; *New York Times* (Oct. 20, 1991), 41, 84.

[585] *Washington Post* (Jan. 22, 1996), A17.

[586] Martha Gil-Montero, *Brazilian Bombshell: The Biography of Carmen Miranda*, (New York: Donald F. Fine, Inc., 1989), 12-18.

[587] Gil-Montero, 20-21.

[588] Gil-Montero, 30.

[589] Gil-Montero, 35-36.

[590] Gil-Montero, 50-56; Robert Sullivan, "Carmen Miranda Loaves (sic) America and Vice Versa," *Sun News* (Nov. 11, 1941); *New York Post*, (Sept. 8, 1941).

[591] Quoted in Gil-Montero, 74 and quoted in Shari Roberts, "'The Lady in the Tutti- Frutti Hat:' Carmen Miranda, a Spectacle of Ethnicity," *Cinema Journal*, 32 (Spring 1993), 8.

[592] Roberts, 3; *New York Times*, (June 20, 1939), 25 and (June 25, 1939), IX, 1.

[593] Quoted in Gil-Montero, 96.

[594] Gil-Montero, 110-119.

[595] *Memo from Darryl F. Zanuck: The Golden Years at Twentieth Century Fox*, Rudy Behlmer, ed., (New York: Grove Press, 1993), 45-46.

[596] Quoted in Gil-Montero, 123.

[597] Roberts, 3; Gil-Montero, 122-124.

[598] Quoted in Gil-Montero, 125, 133-134; "An Artist and Models" in clippings file, Billy Rose Theatre Collection, New York Public Library .

[599] Martin Rubin, *Showstoppers: Busby Berkeley and the Tradition of Spectacle*, (New York: Columbia University Press, 1993), 159-162.

[600] Gil-Montero, 150; Rubin, 167-168.

[601] Quoted in Gil-Montero, 149-150.

[602] Roberts, 15.

[603] Gil-Montero, 153-5.

[604] Gil-Montero, 170, 206, 210; Clippings file, Billy Rose Theatre Collection, New York Public Library.

[605] Gil-Montero, 253-7.

[606] *New York Daily News*, (Oct. 17, 1955).

[607] Roberts, 3, 20.

[608] Gerald Bordman, *American Musical Theatre: A Chronicle*, (New York: Oxford University Press, 1978), 581; *Variety*, (Aug, 12, 1953), (Jan. 27, 1954); Program file, Billy Rose Theatre Collection, New York Public Library for the Performing Arts.

[609] *New York Times*, (Nov. 18, 1951), II, 1 and (Nov. 2, 1951), 19; *Variety*, (Jan. 27, 1954).

[610] "Joey Faye," *The Economist*, (May 10, 1997), 88.

[611] Julian Fox, *Woody: Movies from Manhattan*, Woodstock, N. Y.: Overlook Press, 1996), 53.

[612] Rebecca Ascher-Walsh, "Feast of Stephen," *Entertainment Weekly*, (Feb. 19, 1993), 66.

Chapter 11 Notes: The Musical Banana

[613] S. Frederick Starr, *Bamboula! The Life and Times of Louis Moreau Gottschalk* (New York: Oxford University Press, 1995), 50-51.

[614] Starr, 59.

[615] No one knows with certainty exactly what Chopin said. These are the words of Gottschalk's sister Clara.

[616] Starr, 46.

[617] Starr, 42.

[618] Starr, 75.

[619] Henry Didimus, *Biography of Louis Moreau Gottschalk, the American Pianist and Composer* (Philadelphia: Deacon and Peterson, 1853), 9.

[620] Starr, 81.

[621] Arthur Loesser, *Men, Women, and Pianos: A Social History* (New York: Dover Publications, 1990), 375.

[622] Quoted in Didimus, 8.

[623] Starr, 51.

[624] Didimus, 9-10.

[625] Quoted in Carl E. Lindstrom, "The American Quality in the Music of Louis Moreau Gottschalk," *Musical Quarterly*, 31 (July 1945), 361.

[626] Peter Andrews, "The King of Pianists," *American Heritage*, 34 (December 1982), 67.

[627] Starr, 105.

[628] Andrews, 67.

[629] Starr , 112; Didimus, 19-20; Clyde W. Brockett, "Gottschalk in Madrid: A Tale of Ten Pianos," *Musical Quarterly*, 75 (Fall, 1991), 286.

[630] Starr, 125, 134.

[631] Loesser, 498-499.

[632] Lindstrom, 366.

[633] Quoted in Starr, 153.

[634] Loesser, 511.

[635]Loesser, 500.

[636] Starr, 244-245, 313.

[637] Gottschalk, Louis Moreau, *Notes of a Pianist*, ed. Jeanne Behrend (New York: Alfred A. Knopf, 1964), 39.

[638] Gottschalk, 23.

[639] Gottschalk, 26-27.

[640] Gottschalk, 66.

[641] Starr, 322.

[642] Gottschalk, 102-103.

[643] Gottschalk, 171, 155, 163.

[644] Gottschalk, 99, 261.

[645] Gottschalk, 272.

[646] Starr, 370-374.

[647] Dieter Hildebrandt, *Piano Forte: A Social History of the Piano*, trans. Harriet Goodman (New York: George Braziller, 1988), 118.

[648] Starr, 377.

[649] Gottschalk, 315.

[650] Starr, 380, 384.

[651] Starr, 435

[652] Starr, 184-185, 347.

[653] Lindstrom, 363.

[654] *New York Times*, (May 4, 1979, 64.

[655] Nat Shapiro, ed., *Popular Music, vol. 5, 1920-1929* (Detroit: Gale Research, 1969), 19.

[656] Gregory Koseluk, *Eddie Cantor: A Life in Show Business* (Jefferson, N. C.: McFarland, 1995), 92-93; David Ewen, ed., *American Popular Songs* (New York: Random House , 1966), 453-454; Eddie Cantor with Jane Kesner Ardmore, *Take My Life* (New York: Doubleday, 1957), 165.

[657] *Fruit Dispatch*, (July 1923), 107.

[658] Cantor, 166.

[659] *Detroit News* quoted in *Fruit Dispatch* (August, 1923), 144-145.

[660] George Allan England, "A Once Popular Song Now Sung in China," *Fruit Dispatch* (March 1928), 311.

[661] Ewen, 454.

[662] Quoted in Charles Hamm, *Yesterdays: Popular Songs in America* (New York: W. W. Norton and Co., 1978): 368; *Unifruitco* (May 1930), 624.

[663] Roger Lax and Frederick Smith, *Great Song Thesaurus*, 2nd. ed. (New York: Oxford University Press, 1989), 427.

[664] *Songs of the 1920's*, (Minnesota: Hal Leonard Publishing), 6.

[665] *Fruit Dispatch* (July 1923), 107.

[666] Charles Morrow Wilson, *Empire in Green and Gold: The Story of the American Banana Trade* (New York: Henry Holt, 1947),184.

[667] Will Rogers, "Slipping the Lariat Over," *Fruit Dispatch* (Sept. 1923), 181-183.

[668] Ewen, 454; Shapiro, 113.

[669] Marcia and Jon Pankake, *A Prairie Home Companion Folk Song Book* (New York: Viking, 1988), 94.

[670] *Unifruitco* (Jan. 1948), 9; Wilson, 185.

[671] *Unifruitco* (Jan. 1948), 9-10.

[672] *Unifruitco* (Nov. 1948), 19.

[673] *Unifruitco* (Jan. 1948), 10; "The Chiquita Jingle" at http://www.chiquita.com

[674] Chiquita Brands, Inc., 1999

[675] *New York Times* (Sept. 9, 1993), C6.

[676] Sharon Fitzgerald, "Belafonte the Lionhearted," *American Visions* (Aug.-Sept. 1996), 12.

[677] William Attaway, *Calypso Song Book* (New York: McGraw-Hill Book Company, 1957), 6-7, 9, 12; Arnold Shaw, *Belafonte: An Unauthorized Biography* (Philadelphia: Chilton Co., 1960), 231.

[678] Shaw, 233.

[679] Attaway, 11.

[680] Henry Louis Gates, Jr., "Belafonte's Balancing Act," *The New Yorker*, (Aug. 26, 1996), 134.

[681] Fitzgerald, 12.

[682] Shaw, 229-230.

[683] Quoted in Gates, 138.

[684] Gates, 140.

[685] Adapted from "Thirty Thousand Pounds of Bananas," words and music by Harry Chapin, *Harry Chapin: A Legacy in Song* (Cherry Land Music Company, 1987).

[686] Peter M. Coan, *TAXI - The Harry Chapin Story* (Port Washington, N. Y.: Ashley Books, Inc., 1987), vii.

[687] Coan, 145-146.

[688] Quoted in Coan, 147.

[689] *Los Angeles Times (Ventura County Edition),* (June 18, 1998), B-3.

[690] "Harry Chapin is Gone, But Friends Carry His Song in Their Hearts," *People Weekly* (Dec. 21, 1987), 49.

Chapter 12 Notes: Celebrating the Banana

[691] *New York Times,* (Sept. 6, 1998), Sec. 4, 4.

[692] Kay Martin, interview with the author, Fulton, Kentucky, May 30, 1998.

[693] Elaine Forrester, interview with the author, Fulton, Kentucky, May 29, 1998.

[694] *Fulton Daily Leader*, (Sept. 20, 1982), 10.

[695] Martin interview.

[696] Martin interview.

[697] Forrester interview.

[698] Forrester interview.

[699] *Fulton Daily Leader*, (Sept. 20, 1982), 31.

[700] *1974 International Banana Festival Program*, 16-17.

[701] Carlton J. Corliss, *Main Line of Mid-America: The Story of the Illinois Central*, (New York: Creative Age Press, 1950), 264; *Illinois Central Railroad Company, Organization and Traffic of the Illinois Central System*, (Chicago: Illinois Central Railroad Company, 1938), 11.

[702] Corliss, x; *Illinois Central Railroad Company*, 239, 244-6.

[703] Corliss, 401-7.

[704] *Fulton Daily Leader*, (Sept. 20, 1982), (Sept 21, 1987), 2; *Memphis*, (Aug. 17, 1979), C - 21.

[705] "A History of the International Banana Festival," *International Banana Festival Program - 1971.*

[706] Martin interview.

[707] Forrester interview.

[708] *Fulton County News*, (Aug. 24, 1972).

[709] Martin interview.

[710] *International Banana Festival Program - 1971*, 17, 19; *Fulton Daily Leader*, (Sept. 21, 1987), 9.

[711] Martin interview.

[712] Martin interview.

[713] Ken Bannister, interview with the author, Altadena, California, Feb. 7, 1997.

[714] *Los Angeles Times*, (Feb. 7, 1997), A14.

[715] Bannister interview.

[716] Bannister interview.

[717] Bannister interview.

[718] Bannister interview; *Los Angeles Times*, (June 9, 1998), B-3; Reuters, (March 18, 1998).

[719] Bannister interview.

[720] Bannister interview.

[721] Bannister interview.

[722] *Daily News, L.A. Life*, (Jan. 14, 1998), 10-11; Bannister interview.

[723] Bannister interview.

[724] Bannister interview.

[725] Forrester interview.

[726] Bannister interview.